ON THE FOUNDATIONS OF COMPUTING

On the Foundations of Computing

GIUSEPPE PRIMIERO

OXFORD
UNIVERSITY PRESS

OXFORD
UNIVERSITY PRESS

Great Clarendon Street, Oxford, OX2 6DP,
United Kingdom

Oxford University Press is a department of the University of Oxford.
It furthers the University's objective of excellence in research, scholarship,
and education by publishing worldwide. Oxford is a registered trade mark of
Oxford University Press in the UK and in certain other countries

© Giuseppe Primiero 2020

First Edition published in 2020

Impression: 1

Published in the United States of America by Oxford University Press
198 Madison Avenue, New York, NY 10016, United States of America

British Library Cataloguing in Publication Data
Data available

Library of Congress Control Number: 2019947624

ISBN 978–0–19–883564–6 (hbk.)
ISBN 978–0–19–883565–3 (pbk.)

DOI: 10.1093/oso/9780198835646.001.0001

Printed and bound by
CPI Group (UK) Ltd, Croydon, CR0 4YY

In memoria di mio padre

Preface

The topic of this book is the *Foundations of Computing*. The plural noun *Foundations* refers to the structure of this book, construed around what we argue are the three main foundations of the discipline:

- *The mathematical foundation*. Part I of the volume explores the historical and conceptual background behind the formal understanding of computing, originating at the end of the nineteenth century. It represents the very beginning of what has passed in the literature as the theoretical foundation of computer science, with the formal origins of the notions of computation, algorithm, and program. The topics covered in this first approach to computing are:
  - the definition of effective calculability;
  - the notion of computability by a Turing Machine;
  - algorithms and their nature;
  - the decision problem;
  - the Church-Turing Thesis;
  - the problem of formal verification;
  - the view of computing as a mathematical discipline.

- *The engineering foundation*. Part II of the volume overviews the construction of physical devices to perform automated tasks and associated technical and conceptual issues. We start with the design and construction of the first generation of computing machines, explore their evolution and progress in engineering (for both hardware and software), and investigate their theoretical and conceptual problems. The topics covered in this second part are:
  - technologies from the first to the fourth generation of computers;
  - Moore's Law;
  - programming languages and their paradigms;
  - Lehman and Belady's, Amdahl and Gustafson's Laws;
  - the relation between specification and implementation;
  - correctness and malfunctioning in software.

- *The experimental foundation.* Part III of the volume analyses the methods and principles of experimental sciences founded on computational methods. We study the use of machines to perform scientific tasks, with particular reference to computer models and simulations. The topics covered in this third part are:
  - computational models;
  - computational experiments;
  - computer simulations;
  - controllability and explanation in experimental computing.

Each of these areas is a topic in its own right, and they are often treated separately. Several primers, handbooks, and manuals are available for the technical foundations of computing: computability theory, information theory, computer architecture, formal methods, programming languages, networking, computer models and simulations, and so on. These are essential contributions for an in-depth knowledge of the technical aspects. But none of these works can, by definition, provide a comprehensive, critical, and diachronic overview of the various intersections and relations between those areas: these intersections are, we believe, what have made computing the complex science it is today. Moreover, this volume aims at offering a critical, i.e. philosophical, analysis of the conceptual issues that computing presents across its different foundations. The understanding of these relations is an essential need for students and researchers across several disciplines, most notably computer science, philosophy, science and technology studies. We hope to have provided satisfactory bibliographical references at the end of this volume to illustrate the wealth of studies and results that each single area on its own has formulated over the decades. In choosing to present the foundations as a unity, we claim to offer a more complete and satisfactory view of the different aspects and associated difficulties in understanding the nature of computing.

The second element that requires some qualification in the title of this volume is the noun *computing*: by it we refer inclusively to the abstract and physical aspects of computational processes, the formal and experimental nature of the use of computing techniques in the formation of scientific results, and the logical and material limits of processes based on computational aims. Hence, our focus in this volume is not restricted to computability theory. It also does not exclude the analysis of structural properties of computational systems and their limitations. In this sense, *computing* encompasses all the various forms in which the discipline manifests itself in its academic incarnation (computer science) but also in its industrial form (business computing) and its daily social phenomenon (information technologies). More importantly, we intend *computing* to be considered as a discipline of theoretical, technical, historical, and epistemological investigations. In this broader meaning, the *Foundations of Computing* have the philosophical presumption of providing a conceptual analysis of some of the most important (although not all) issues in computing.

This variety of aspects, which this book intends to connect, is the result of a historical evolution and it reflects the current status of a discipline of crucial importance for science and society at large. Computing, today more than ever before, is a multi-faceted discipline which collates several methodologies, areas of interest, and approaches: mathematics, engineering, programming, applications. Accordingly, computer science has been variously defined as: a mathematical discipline, an engineering science, a science of information, the study of computing machines, a practical and experimental science. Given its enormous impact on everyday life, it is essential that its debated origins are understood, and that its different foundations are explained. The aim of this volume is to offer a comprehensive and critical overview of the birth and evolution of computing, and to analyse some of the most important technical results and philosophical problems of the discipline.

From the methodological point of view, this volume combines both historical and systematic analyses. The history of computing offers an essential method to understand the evolution of the discipline: although not a work of historical research, the help of many colleagues have made a strong impact on this volume, which attempts at preserving at least a minimum of historical awareness concerning the multiplicity of actors and trends on whose basis computing evolves. The philosophy of computing complements the diachronic presentation of issues by providing a much-needed interpretation of the many methodological, ontological, and epistemological issues at its core: in this task, a formal understanding of the discipline expressed by logical methods emerges, completed by an analysis of socio-technical factors. For computer science practitioners, a foundational, historical, and philosophical approach becomes essential to understanding the past of the discipline, and to figure out the challenges of the future. Students and researchers (both of computer science and of the history and philosophy of computing) are in need of a tool that combines the technical aspects of the foundations of computing with its historical reconstruction and philosophical analysis. Unfortunately, too often historical and philosophical analyses remain disconnected from the technical and applied aspects of the science they are supposed to illuminate. This volume hopes to fill this gap.

The structure of the volume allows readers to select the chapters of interest at will, depending on the context and audience. As a textbook, this volume covers the essential material for undergraduate courses on several foundational aspects of computing:

- Computability theory, Chapters 2 to 5;
- Computer architectures, Chapters 8 and 9;
- Computer modelling and simulation, Chapters 13 and 14.

This volume is also a scholarly contribution to the growing area of history and philosophy of computing. The debates it surveys are among the latest and most urging ones:

- The crisis of foundations in mathematics and the birth of the decision problem, Chapter 1;
- What are algorithms, Chapter 6;
- The debates on computational artefacts and malfunctioning, Chapters 10 and 11;
- The analysis of simulations as computational experiments, Chapter 15.

By covering these topics, the volume provides a much-needed resource to contextualize the foundational issues and it works as a reference to further literature. It contributes to frame our research in the appropriate conceptual space.

Finally, each part of this volume guides the reader through results and theories and concludes by offering an analysis of the notion of computational validity in the light of the corresponding foundational reading:

- Formal computational validity, Chapter 7;
- Physical computational validity, Chapter 12;
- Experimental computational validity, Chapter 16.

Each of these chapters provides a new understanding of the criteria according to which a computational process can be considered valid. In order to do this, we also provide a larger contextual analysis by which: the notion of algorithm is redefined; the cases of error in physical computing are analysed; and the criteria for validation and verification for computational experiments and models are presented.

These results rely crucially on the method of levels of abstraction. We apply this to define computational systems as abstract, physical, and experimental artefacts. Identifying the different levels of abstraction at which computing and its related systems can be analysed is crucial to defining their formal properties, to identifying the corresponding material laws and limits, and to expressing the epistemological characteristics of computing artefacts in their scientific uses. This methodological aspect is recurrent throughout the present investigation and it is particularly important in our definitions of computational validity, where appropriate variables of interest are identified: language and encoding for the formal version; architecture and behaviour for the physical version; realization of the formal structure in the physical implementation, and relation with models for the experimental version.

Each chapter of the book starts with a short summary meant to highlight its main contribution and to facilitate the reader in connecting the dots among the different historical and conceptual moments explored by the book. The second feature present in each chapter is a list of exercises, which have a double aim: for the reader they offer a lightweight test to verify that the main topics of each chapter have been absorbed; for the teacher who would wish to use this volume as a didactic aid, they offer possible tests or a guide to examination. In short, this volume is both a handbook for students of courses in philosophy of computing (with a historical characterization of the topics presented) and a tool for researchers who wish to have a comprehensive overview of issues in this area, both technically and formally introduced, and who can use it as a wedge towards further literature.

This volume is obviously presented as a stand-alone investigation of the foundations of computing, and it is written with the intention of being completely self-contained. Nonetheless, as the author of this volume, I also feel compelled to say a few words on how this volume fits within the larger scheme of other scientific and academic writings for which I am responsible. The logical root of computing and the decision problem in the classical and intuitionistic approach represent topics at the very beginning of my logical and philosophical investigations, since my early interests in the epistemological analysis of constructive truth and the possible weakening offered by the notion of information (see Primiero (2006, 2009, 2013)). The problem of formal computational validity as a variant to the notion of logical validity, which emerges in the first part of the volume, has interested me for some years (see Primiero (2015)). Logical methods for the identification of properties in computational artefacts has been a recurrent theme in my research until today (see Primiero (2017); Angius and Primiero (2018)). The problem of miscomputation and software malfunctioning, which I have explored with co-authors (see Fresco and Primiero (2013); Primiero (2014); Floridi et al. (2015); Primiero et al. (2018b)) is at the basis of the issue of physical computational validity investigated in the second part of this volume. The experimental interpretation of computational validity, explored in the third part of the volume, is the latest step in this progression, whose first results are included in this volume and are based on conceptual and experimental work related to the computer simulation of multi-agent systems (see Battistelli and Primiero (2017); Primiero et al. (2017a,b); Bottone et al. (2018); Primiero et al. (2018a,c); Primiero (2019b)). From a larger conceptual perspective, the foundations of computing as interpreted in this book fall within the research area of the philosophy of computing, a growing field of investigation which covers the large variety of topics touched upon in this volume (De Mol and Primiero (2014, 2015); Primiero (2016)).

The present attempt is not a fully comprehensive analysis of all the interesting foundational issues in computing. We only focus on the basics of classical computation, without extending its scope to any of the more recent trends (e.g. quantum or infinite computations); it only considers data processing up to parallelism, without entering the vast and deep field of big data; it only analyses scientific computing in relation to models and simulations from a theoretical viewpoint, without aiming at covering all the basics of high-performance computing. Moreover, we do not enter important debates in the current literature, like the ethics and politics of artificial intelligence, machine learning, and its applications, which obviously constitute an important aspect of today's foundational view of computing. Our task here is more pressing in reformulating and contextualizing those foundations common to all current and future forms of computational processes. Also, ours is not the only attempt in this direction. The scientific community has few resources that aim at covering—at least partially—the breadth and variety of topics proposed by this volume. But none of them offers the combination of technical, historical, and philosophical aspects present here. The main tenet of the present book, that computing is underpinned by three distinct and complementary paradigms which we argue are all foundational, was argued to some extent in Eden (2007), where the three approaches are identified as the rationalist, the technocratic, and the scientific ones. In this very same direction, this book is closest in spirit to Tedre (2015), which is currently the only volume on the market which tries to capture the breadth and complexity of computing foundations. But it is largely discursive and entirely non-technical, nor formal. The present contribution, on the other hand, insists on the need to combine conceptual, technical, and formal analyses required for a deep understanding of the foundations of computing at large. Moreover, we present current topics in the philosophy of computer science, for example, the debate on the nature of algorithms, or the definition of malfunctioning for computational artefacts. We connect each of the older debates in one of the foundational areas of computing with the most recent philosophical analyses. In this light, Turner (2018) is the most recent contribution in the direction of formulating a formal and technical analysis of the issues at the basis of the philosophy of computer science. The present volume differs from it both for the larger context in which we present some of the same topics, and also for the elements we consider essential to such analysis, including validity, malfunctioning, and the relation between models and simulations. Similarly, Rapaport (2018) has for many years established a free resource on the Internet for a variety of topics converging around the relation between computing and philosophy: while we do not aim at offering a comparably extensive presentation of topics, we do hope to have formulated a consistent and coherent analysis around the different foundations of computing and provided a deep basis on which to explore further problems. Davis (2012) is a well-informed and extremely pleasant contribution to the literature, which covers the origins of automatic reasoning from Leibniz and ends with Turing. It offers a historical overview completed by technical details, further extended in footnotes. It is another brilliant example of scientific literature accessible to the general public. From the point of view of this volume, it does not extend the historical analysis beyond the mathematical foundation, and as such does not touch on either the engineering aspects of the discipline, nor on the experimental one, it does not link the mathematical foundation to programming theory and practice, and it does not investigate the philosophy of computer science debates. Similarly, Priestley (2011) covers the early development of automatic computing, the theoretical work of mathematical logicians such as Kleene, Church, Post, and Turing, and the machines built by Zuse and Aiken in the 1930s and 1940s. It provides a rather technical discussion of the role of logic in the development of the stored program computer. From the point of view of the present proposal, it does not extend the historical analysis beyond the first generation of computers and the analysis of programming paradigms is limited to the more logic-based ones, especially ALGOL. In doing so, it does not give the reader a sense of continuity with what are now acknowledged as the main programming paradigms, nor modern computer architectures and their problems, it does not offer a philosophical analysis of problems such as the interpretation of algorithms, formal verification, and malfunctioning, and it does not investigate a new major view on computing, namely as an experimental science. Daylight (2012) focuses on Turing's legacy on software engineering, and as such it overlaps in spirit with the present proposal as far as determining the mathematical roots of computing is concerned. It also provides an extensive analysis of its influence in the context of very specific figures in the history of software engineering. It clearly has a more historical focus than what is aimed at by the present volume, as illustrated by the oral history interviews it includes. But the volume is largely discursive and only some formal aspects are provided, it does not cover philosophical problems in computer science, it does not focus on the laws of software engineering, and it does not approach the experimental foundations of computing. A number of other more specific volumes could be mentioned: Adams (2011) for a detailed historical and technical analysis of symbolic logic and the origins of recursive function theory; O'Regan (2012) for the history of Computing in general;

Boolos et al. (2002) for Computability Theory; Morrison (2015) for the computational problems in the philosophy of science.

This quick overview is meant to acknowledge the large number of contributions within which the present volume is situated, but also to stress that the intersection of historical, philosophical, and technical roots of computing which we investigate is novel in spirit and aim.

Acknowledgements

A s with any long and complex project, help, encouragement, suggestions, and strength
come in many forms and by many people.

This particular book reflects the years I spent as a member of staff in three different academic institutions (the Center for Logic and Philosophy of Science at Ghent
University, the Department of Computer Science at Middlesex University London, and
the Department of Philosophy at the University of Milan) and as a visiting fellow in
several others (Oxford University, City University of New York, Newton Institute for
Mathematical Sciences, University of Stockholm, Politecnico di Milano, Università degli
Studi di Sassari). In all these places, I held courses or lectures on topics which would
eventually become parts of this volume. It is not the case that at every point of this journey
I was aware of how that experience was contributing to the final product that is this book,
but now I know this to be the case. I wish to thank all the colleagues (too many to name
them all) and students whose path I have crossed and who—knowingly or not—have
helped me with critiques, comments, requests of clarification.

The life of an academic is built around the courses they give and the meetings they
have with others to exchange ideas. Of the several conferences and workshops I have
attended in these years, I need to mention in particular: the Conferences and Symposia
of the Commission for the History and Philosophy of Computing (HaPoC, www.hapoc.
org); the Conferences of the International Association for Computing and Philosophy (IACAP, www.iacap.org); the Workshops on the Philosophy of Information (SPI,
https://socphilinfo.github.io/); the Annual Conference of the Association Computability in Europe (ACiE, www.acie.eu). Since 2017 the research efforts in the history and philosophy of computing have received a most welcome support through the meetings of the
ANR research project *What is a program? Historical and philosophical perspectives* (https://
programme.hypotheses.org/) where some parts of this book have been presented.

The following people need to be mentioned explicitly, for their professional help
and friendship: Patrick Allo, Nicola Angius, Selmer Bringsjord, Felice Cardone, Barry S.
Cooper (†), Marcello D'Agostino, Martin Davis, Edgar Daylight, Liesbeth De Mol, Juan
M. Duran, Luciano Floridi, Nir Fresco, Nikos Gorogiannis, Hykel Hosni, Bjørn Jespersen,
Per Martin-Löf, Simone Martini, Elisabetta Mori, Tomas Petricek, Mark Priestley, Franco
Raimondi, Viola Schiaffonati, Wilfried Sieg, Göran Sundholm, Jacopo Tagliabue, Matti
Tedre, Raymond Turner, Franck Varenne, Mario Verdicchio, and all my co-authors.

Along with colleagues, many others have been close to me during the years, and in particular during a difficult period in which all the energies left have been put to realize this book. I wish to thank Mirjam, for her love. My family, for what has been and what will still be. My friends, for being there.

London
Palermo
Milano
April 2019.

Contents

1 Introduction

How Algorithms Control the World. (*The Guardian*, 1 July 2013).
Rise of the Machines. (*The Economist*, 4 April 2018).
Are We Living in a Computer Simulation? (*Scientific American*, 7 April 2016).

These are real headlines appearing on major journalistic outlets during the last few years. They indicate a recurrent and strengthening feeling in the relation between humanity and computing technologies: fear. In the first case, it is fear of opaque structures, faster than we can ever be, and decisive in all aspects of modern life. In the second case, it is fear of complex artefacts, whose reliability is crucial in several daily events in which our lives can be put at stake by not so uncommon errors. In the third case, it is the atavistic fear of our origin and destiny, rephrased and redesigned around the myth of a computational mastermind.

As often with fears that pervade human culture and science in particular, their origins lie in lack of knowledge. The wealth of popular science publications that investigate the effects of computing technologies on work, health, economy, and social life tend to assume one of two tendencies: on the one hand, the enthusiasts who share their visionary future where algorithms and machines will be our companions, friends, and saviours, lifting the burden of hard work and elevating us from the last remnants of our mortal lives; on the other hand, the detractors who paint a bleak vision of malevolent AIs fighting humanity for their right to survive.

Science and its philosophical analysis have in this context a double role: to reduce ignorance on which extremist views grow, and to build the basis for a more realistic, mature, solid, and argumentative position on our current and future relation with machines and computing technologies.

This aim requires a difficult but essential combination of factors. First, computing technologies need to be explained, to all sorts of specialists as well to the general public, to reduce the current algorithmic fever. This means to illustrate the basis of the mathematical and engineering tools that have for decades built up towards the creation of computing machines, and to explain the laws that regulate their progress. Second, everyone,

On the Foundations of Computing. Giuseppe Primiero, Oxford University Press (2020). © Giuseppe Primiero.
DOI: 10.1093/oso/9780198835646.001.0001

and scientists in particular, need to ask the difficult questions concerning technology: how do we define algorithms, and which properties do we ascribe to them? how is a computational error recognized and its negative effects constrained?; what does it mean to know computationally? These questions may appear trivial or too generic at first, but their answers constitute the backbone for any strategy, politics, or ethics involving computing technologies. In turn, this can determine our way of living with them. For this task, the sole scientific approach becomes insufficient. And a pure philosophical analysis risks being too pretentious, presupposing (or worse: ignoring) the technical basics. A combination of solid scientific grounding and consistent philosophical research can lead to an understanding of how computing is affecting our epistemological and ontological worldviews.

Our first task is therefore to locate the origin of the current frenzy in its appropriate historical and formal context: the problem of deciding mechanically which sentences of a mathematical system are true. This problem was sought to be resolved through the definition of computable functions and in turn by modelling this latter idea in the form of a universal machine capable of realizing all such functions. The notion of algorithm and its execution is thus recognized in the mechanical execution of a well-defined function, a concept which matches the notion of program. The connection between programs and proofs (as the standard formal way of determining the truth of propositions in mathematics) is thus at the very heart of the notion of computation. It is nonetheless a lot more complex to establish which properties can be ascribed and should be defining algorithms. It is striking to witness the superficiality and confusion shown by those who do not deal with computing professionally. But it is also amazing to realize that a profound and intense debate concerning the nature of algorithms and their implementations is still ongoing among computing professionals. This points to the difficulty of analysing objects whose nature is at the same time both formal and concrete, abstract and linguistic. Along with their definition, the formal analysis of their correctness becomes essential, a guarantee *in principle* that their intended behaviour is satisfied.

The second task is to highlight the parallel development of machines capable of realizing those same functions identified as computable, a step from the abstract realm of mathematics to the concrete one of engineering. It is in this context that the limitations to (formal) correctness as well as the new laws guiding the evolution of computing machinery emerge, to define a new concept of validity. While the practice of automated testing and formal verification constitute today an essential aspect of controlling computing machinery especially in safety-critical systems, large web-based, robotic, and AI applications, there is still an extensive conceptual void to be filled. It concerns our ability to define, categorize, and explain the limits of functionality, usability, and efficiency for *implemented* programs, i.e. in the passage from formal to physical correctness. Specification (as the expression of intention) and implementation are required to reflect the complex, stratified nature of computing machinery.

A third task consists in clarifying in which way computing machinery can help forming our scientific view of the world, and under which conditions results obtained through computational means can be granted scientific validity. This is an essential step for two reasons: first, more and more of our science is the result of using computing technologies extensively; second, from the previous analysis we should be aware of the limits induced by mechanical processes which are in principle grounded on a formal notion of validity, but whose realization is based on implemented processes with physical and contextual properties. In this sense, the mentioned problem of recognizing our reality as the possible result of a computation is more constructively and realistically rephrased as the task of identifying the characteristics that make a simulation a valid way of knowing any external reality. In this third task, the power and the limits of computational methods need to be fully formulated and qualified to obtain a transparent understanding of computing and its properties.

These three tasks reflect a way to concretely understand computing and its realizations. In doing so, the myths pervading our information society can be dispelled. More interestingly, we can contribute to the important and essential job of building the philosophical analysis required by computation and associated methods, a feature so intrinsic to our culture today that it is hard to imagine a more urgent task for philosophy. A solid and critical understanding of the foundations of computing is a new and essential building block in the relation between society and technology, a task that philosophy and science need to share and act upon urgently.

The present volume is our contribution towards a more balanced scientific understanding of the new information society.

Part I
The Mathematical Foundation

2 A Fundamental Crisis

Summary

This chapter reconstructs the historical background of the mathematical foundation of modern computing. At the beginning of the twentieth century, mathematical correctness and truth were under scrutiny and three main theoretical positions emerged in trying to offer an appropriate methodological answer to their definitions. The resulting debate was the basis for explaining what it means to compute.

2.1 The Foundations of Mathematics Debated

While computing technologies are young by comparison with other efforts of human ingenuity, their origins go back to many early chapters in the human quest for the understanding and realization of mechanical aids to knowledge. An exhaustive history would need to reconstruct the influences of the philosophical views on the mechanization of deductive knowledge in the *Ars Magna* by Ramon Llull from the thirteenth century; in the *Characteristica Universalis* and *Calculus Ratiocinator* by Leibniz in the eighteenth century; and the (only partly realized) engineering efforts for the *Analytical Engine* by Babbage in the nineteenth century.[1] But while philosophy and engineering have been crucial disciplines in the birth and evolution of computing, largely acknowledged in the following part of this book, our recollection of the foundations of computing starts from their roots in mathematics.

---

[1] See for example Davis (2012); O'Regan (2012); Tedre (2015) for short recounts of these contributions to computing. For references to the work of Llull, see http://orbita.bib.ub.es/llull/ and Llull (1985). For Leibniz' contribution to the idea of a universal language, see e.g. his *On the General Characteristic* (ca. 1679) in (Leibniz, 1989, pp.221–8). For the works of Babbage, see Babbage (2010).

On the Foundations of Computing. Giuseppe Primiero, Oxford University Press (2020). © Giuseppe Primiero.
DOI: 10.1093/oso/9780198835646.001.0001

Between the end of the nineteenth and the beginning of the twentieth century, a foundational crisis was taking place across the most important mathematical circles in the world. This crisis can be considered (factually and metaphorically) as the root of computing in mathematics. From the ruins left by the titanic clash of different views on mathematical truth, infinity and correctness, the *theory of computation* was born. This term refers today to the field of logical and mathematical research that investigates the concept of *effective computation,* or less abstractly, of *effectively computable mathematical structures.* This discipline, which expresses the essential formal aspect of computing, should be considered in its coexistence with the theory of algorithms, whose origins actually pre-date the theory of computation. Algorithms will be analysed later in their philosophical, mathematical, and programming interpretations. This interplay of ideas, notions and scientific research programmes is a constant in our conceptual reconstruction of computing as a discipline, and should be always recalled by the reader.

In the following part of this chapter, we will offer a reconstruction of the debate on the foundations of mathematics that was at the origin of the notion of computable function and, in turn, of the very idea of computing as a mathematical discipline.

2.2 Logical Roots

Computability theory as the mathematical roots of computing has its own origins in logic. In its most basic formulation, logic studies how to perform correct reasoning. The correctness of this abstract act of knowledge is formally made concrete in the form of two relations between sets of sentences: on the one hand, by the definition of valid *consequence*; on the other, by that of correct *derivation*. These two notions express the current view on, respectively, the semantic and syntactic interpretations of logical reasoning.

The notion of consequence is based on the interpretation of propositions as truth-bearers, expressing (the obtaining of) corresponding states of affairs. This tradition has been codified in the work of Tarski,[2] which in turn goes back to Quine, Frege, and Bolzano. Truth as correspondence relies on the notions of *interpretation* and of *model,* in which truth is realized. While the basic intuition of truth for the propositional translation of sentences only requires an assignment of truth values (true/false) to atoms and their closure under logical connectives, for the aim of developing formal systems in various branches of mathematics the use of predicative calculi was essential, and so the definition of an interpretation for predicative formulae:

Definition 1 (Interpretation) *An interpretation of a formula φ is an assignment of meanings to any individual variable x of φ or predicative variable P(x) of φ for objects and predicates (or relations for predicates ranging over more than one variable) in the language of a system F.*

[2] Tarski (1943).

As for the propositional case, predicative formulae interpretation is completed by the definition of compositional rules for logical connectives and the standard extension to the universal and existential quantifiers. The notion of truth by interpretation is used to define the concept of model:

Definition 2 (Model) *A model $\mathcal{M}$ of a sentence ϕ (or a set of sentences Γ) is an interpretation in which ϕ (or every ϕ member of Γ) is true.*

The semantic notion of consequence uses preservation of truth in a model:

Definition 3 (Semantic Consequence Relation) *Given a set of sentences $\Gamma = \{\phi_1, \ldots, \phi_n\}$ within some formal system F, we say that a formula ψ is a semantic consequence of Γ in F, denoted as $\Gamma \vDash_F \psi$ if and only if there is no model $\mathcal{M}$ of F such that all members of Γ are true in $\mathcal{M}$, but ψ is false in $\mathcal{M}$.*

A different understanding of the notion of correct logical reasoning comes from the syntactic reading of the relation between sentences. Under this reading, the truth of a formula ψ is reduced to its correct derivation by means of a proof procedure as a relation between sentences, the premises, and the conclusion, i.e. the formula ψ at hand. The notion of *proof procedure* (or derivation) therefore grounds the syntactic understanding of correct reasoning:

Definition 4 (Proof Procedure) *A proof procedure $\mathcal{P}$ of a sentence ψ from a set of sentences $\Gamma = \{\phi_1, \ldots, \phi_n\}$ is a finite set of steps in which every formula is either some ϕ_i member of Γ, or is obtained by a previous step by applying a well-defined and explicit rule r of a formal system F, and the last step in $\mathcal{P}$ is the formula ψ.*

With this notion we can define the syntactic notion of derivability:

Definition 5 (Syntactic Derivability Relation) *Given a set of sentences $\Gamma = \{\phi_1, \ldots, \phi_n\}$ within some formal system F, we say that a formula ψ is syntactically derived by Γ in F, denoted as $\Gamma \vdash_F \phi$ if and only if there is a proof procedure $\mathcal{P}$ in F of ψ from Γ.*

When the set Γ of sentences of interest represents some scientific theory (like axioms and laws of geometry, physics, or arithmetic), semantic consequence and syntactic derivability relations tell us what are the truths, respectively theorems, of that theory. Towards the end of the nineteenth century, and during the first decades of the twentieth, mathematicians were interested in knowing how much could be proved by their theories, i.e. whether they were strong enough to validate all the truths of their respective fields. More clearly stated, the main research problem for mathematicians at that time can be formulated as follows:

Problem 1 (Soundness and Completeness) *Can we build logical systems that allow us to discover all (completeness) and only (soundness) the true sentences derivable from the axioms of a given theory?*

It was the main aim of mathematicians to ensure that their designed logical systems would guarantee sound and complete theories. This was true in particular of arithmetic.

The uncertainty surrounding this crucial issue, and the related methodological problems, became known as the *Grundlangenkrisis*, or the crisis of the foundations of mathematics.

One historical event can be chosen as representative of this situation. The *Second Conference on Epistemology of the Exact Sciences* took place between 5 and 7 September 1930 in Königsberg: it can be considered a turning point in the history of the philosophy of mathematics and logic. Three lectures were delivered during that conference, conveying the basic research programmes in the foundations of mathematics:[3]

1. Rudolf Carnap, *Die logizistische Grundlegung der Mathematik;*
2. John von Neumann, *Die formalistische Grundlegung der Mathematik;*
3. Arend Heyting, *Die intuizionistiche Grundlegung der Mathematik.*

Each of these lectures illustrated one main position in the debate on the foundations of mathematics: the logicist position by Gottlob Frege with his *Grundgesetze der Arithmetik;* the formalist one by David Hilbert with *Grundlagen der Geometrie;* and the intuionist approach by J.E. Brouwer with *Over de grondslagen der wiskunde.*[4]

2.3 Logicism

Frege's work had been inspired by the idea of grounding mathematics (and arithmetic as a part of it) entirely on logic:

> The most solid way of derivation is clearly that of pure logic which, being abstracted from the particular properties of things is based only on laws, from which all of knowledge comes from.[5]

This programme relied on two main theses:

Thesis 1 (Logical definability) *The concepts of mathematics can be derived from logical concepts through explicit definitions.*

Thesis 2 (Logical derivability) *The theorems of mathematics can be derived from logical axioms through purely logical deduction.*

As a simple example, consider the notion of number, which according to Frege can be explained as the extension of a set: the number 0 is the extension of all concepts that are expressed by empty sets; if you have to explain what 0 means, you can point to all the sets that have zero elements and say that 0 denotes all such sets; the number 1 is the extension of all concepts that are expressed by sets that have just one element;

---

[3] See respectively Carnap (1931), von Neumann (1983), and Heyting (1983). For an overview of the conference, see Reichenbach and Cohen (1978).

[4] See respectively Frege (1903), Hilbert (1902) and, Brouwer (1907).

[5] (Frege, 1967, ch.1).

and similarly for all other numbers. According to this intuition, the function that selects all the sets of a given cardinality can then be used to determine the extension of the corresponding natural number, i.e. which sets fall under it. Starting from this definition of natural number as extension of concepts with the same cardinality, Frege aims at showing that all arithmetical expressions can be logically derived from other more simple notions. For its universal reliance on logic, Frege's programme became known as *Logicism*.

Notoriously, Frege's programme was crushed by the discovery that one could derive a *paradox* from one of the system's laws in his *Grundgesetze der Arithmetik*. The law in question is formulated as follows:

Definition 6 (Basic Law V) *Given a function $f(x)$, let its extension be $\{x \mid f(x)\}$, and given a function $g(x)$ let its extension be $\{x \mid g(x)\}$; then it holds $f = g$ iff $\forall x(f(x) \leftrightarrow g(x))$.*

If one takes the set of elements x that satisfy the function f, and this set has the same cardinality of the set of elements x that satisfy another function g, then the functions f and g must be extensionally equal, i.e. express concepts identifying the same natural numbers. If one allows f and g to be themselves objects of selection according to cardinality, a paradox famously illustrated by Bertrand Russell in a letter he wrote to Frege in 1902 follows:

You state that a function, too, can act as the indeterminate element [i.e. it can apply to another function]. This I formerly believed, but now this view seems doubtful to me because of the following contradiction. Let *w* be the predicate: to be a predicate that cannot be predicated of itself. Can *w* be predicated of itself? from each answer its opposite follows. Therefore we must conclude that *w* is not a predicate. Likewise there is no class (as a totality) of those classes which, each taken as a totality, do not belong to themselves. From this I conclude that under certain circumstances a definable collection does not form a totality.[6]

The paradox introduced by Russell is equivalent to the following more informal version:

Proposition 1 (The Barber Paradox) *Suppose there is a town with just one barber, who is male. The barber is a man in town who shaves all those, and only those, men in town who do not shave themselves. In this town, every man keeps himself clean-shaven, and he does so by doing exactly one of two things: shaving himself; or being shaved by the barber. From this, one can ask who shaves the barber. This results in a paradox: the barber can either shave himself, or go to the barber; however, neither of these possibilities are valid. If he shaves himself, then he contradicts his own definition; and if he goes to the barber, then he shaves himself again.*

To see how a paradox follows from the theory of extensions of sets, let us reconsider Russell's argument. Define R as the set of all sets that do not contain themselves. This is the equivalent of the barber's definition as the man who shaves all who do not shave

[6] Russell (1965).

themselves. Then one might ask the question whether R is contained in the set R. This corresponds to asking whether the barber shaves himself, or goes to the barber. If the equivalence holds, then by substitution R is a member of R *ad infinitum*, and the set R uses itself to be defined, a *vicious circle*. If R is not equivalent to R, then for every element X of R, it must hold X is not equivalent to R. In this way, Russell showed that an unrestricted definition of set would lead to contradictions.

This paradox represented the greatest problem for the Fregean ideal of a logic able to provide all principles and rules on which to define all mathematical concepts. This idea of logic referred to the formulation of *one* theory and *one* model of interpretation, such that it would provide a unified axiomatization for arithmetic and all the other sciences. Moreover, the problem was not confined to Frege's system, as similar paradoxes were arising in all branches of mathematics. For a different example, let us consider the paradox by Burali-Forti in naive set theory.[7] It is possible to order the elements of infinite sets by using ordinals numbers:

Definition 7 (Ordinals) *The first ordinal is $\varnothing$ or 0. The successor of some ordinal α is defined as $\alpha \cup \{\alpha\}$, or $\alpha + 1$. The union of all members of a set of ordinals is also an ordinal.*

Suppose that α and β are two ordinals, then $\alpha \leq \beta$ if and only if α is a member of β or α is equal to β and it can be proved that there exists a total order relation $\leq$ on the set of ordinals.[8] Let O be the set of all ordinals. It follows from the definition that

$$\varnothing = 0 \text{ is a member of O}$$
$$\varnothing \cup \{\varnothing\} = \{\varnothing\} = 1 \text{ is a member of O}$$
$$\{\varnothing\} \cup \{\{\varnothing\}\} = \{\varnothing, \{\varnothing\}\} = 2 \text{ is a member of O}$$
$$\{\varnothing, \{\varnothing\}\} \cup \{\{\varnothing, \{\varnothing\}\}\} = \{\varnothing, \{\varnothing\}, \{\varnothing, \{\varnothing\}\}\} = 3 \text{ is a member of O}$$
$$\vdots$$
$$\{1, 2, 3, 4, \ldots\} = \omega \text{ is a member of O}$$
$$\omega \cup \{\omega\} = \{\omega, 0, 1, 2, \ldots\} = \omega + 1 \text{ is a member of O}$$
$$\omega + 2 \text{ is a member of O}$$

Definition 8 *The set of all countable ordinals is*

$$\omega_1 = \omega \cup \{\omega, \omega + 1, \omega + 2, \ldots\} \text{ is a member of O}$$

Definition 9 *The set of all ordinals that have the maximal cardinality of the real numbers (or the set of all countable and $\aleph_1$-ordinals) is ω_2, a member of O.*

Consider now the following definition

---

[7] Burali-Forti (1897).

[8] Recall that an order relation $\leq$ is total over a set X if and only if $\leq$ is reflexive, transitive, antisymmetric and for all α, β members of X, it holds $\alpha \leq \beta$ or $\beta \leq \alpha$.

Definition 10 (Totally Well-Ordered Set) *A totally ordered set is well ordered if and only if every non-empty subset has a least element.*

Every well ordered set is associated with a so-called order type. Two well ordered sets *A* and *B* are said to have the same order type if and only if they are order-isomorphic. Every order type is identical to exactly one ordinal.

Definition 11 (Cardinal Number) *The cardinal number of a set A is the smallest ordinal α such that α and A are equinumerous.*

The paradox affecting the theory of ordinal numbers is formulated as follows:

Definition 12 (Burali-Forti Paradox) *In the theory of transfinite ordinal numbers every well-ordered set has a unique ordinal number; every segment of ordinals (i.e. any set of ordinals arranged in natural order which contains all the predecessors of each of its elements) has an ordinal number which is greater than any ordinal in the segment. Assuming by contradiction that the class Ω of all ordinals could be linearly ordered, Ω carries all properties of an ordinal number, namely it would be itself well-ordered and it would possess an ordinal ω member of Ω. Thus Ω would be order-isomorphic to a proper initial segment of itself, the one determined by ω. Then, we can construct its successor $\omega + 1$, which is strictly greater than Ω. However, this ordinal number must be an element of Ω since Ω contains all ordinal numbers, and we arrive at $\Omega < \Omega + 1 \leq \Omega$.*

The definition of set in Frege's paradox and the definition of the class of ordinals in Burali-Forti's paradox both rely on the notion of totality of a collection. Poincaré was the first to recognize the Vicious Circle Principle as the basis of what is known as an *impredicative definition*, i.e. a definition in which a member of a set is defined in a way that presupposes the entire set to be defined already.[9] According to the French epistemologist, to avoid being defined circularly, a mathematical entity was to be defined inductively only on the basis of constructions from starting given objects. Also Russell aimed at restricting the formulation of such totalities, through the analysis of the Vicious Circle Principle:

> If, provided a certain collection had a total, it would have members only definable in terms of that total, then the said collection has no total.[10]

The Russellian solution to the paradox in Frege's system is notoriously formulated within the Theory of Types. The definitional procedure which avoids the vicious circle requires to stratify the objects of predication within the theory so that a predicate can never be applied at the same level of an object for which it is defined. The ramified version of the Theory of Types went beyond this restriction:[11] it allowed non-strictly predicative definitions through the formulation of an axiom to reduce any propositional function to a predicative one.

---

[9] Poincaré (1909).
[10] (Russell, 1903, Appendix B), (Whitehead and Russell, 1962, vol.1, Introduction, ch. 2, p.1).
[11] Russell (1956).

Definition 13 (Axiom of Reducibility) *Given any function ϕ, there is exactly one predicative function μ equivalent to ϕ for any value of the variable x:*

$$\vdash \exists \mu \forall x (\phi x \leftrightarrow \mu! x).$$

The Axiom of Reducibility expresses the principle according to which every definable function ϕ is correctly applied to an object x if and only if it has exactly one predicative formulation μ for each object x to which it applies. The existential claim presupposed by the formulation of μ was the main point of critique in the epistemological value of this axiom. Two further axioms in use in the *Principia* raised doubts because of their existential nature. The Axiom of Choice was formulated as follows:

Definition 14 (Axiom of Choice) *Let C be a collection of non-empty sets. Then there exists a function f defined on C with the property that, for each set S member of C, $f(S)$ is a member of S.*

By the Axiom of Choice, for every well-defined collection of non-empty sets there is one function selecting an element from each set in the collection. This is equivalent to saying that the Cartesian product of a collection of non-empty sets is always non-empty. The second relevant axiom is formulated as follows:

Definition 15 (Axiom of Infinity) *There is a set I (the infinite set) such that $\varnothing$ is a member of I and such that for every x member of I it holds that $(x \cup \{x\})$, i.e. the union of x with the singleton set containing x is a member of I.*

Consider a standard definition of natural numbers in terms of sets, for example the recursive definition of natural numbers used in Zermelo-Fraenkel set theory:[12]

$$0 = \varnothing$$
$$1 = 0' = \{0\}$$
$$2 = 1' = \{0,1\}$$
$$3 = 2' = \{0,1,2\}$$
$$\vdots$$

Then the Axiom of Infinity asserts the existence of all natural numbers and by the Axiom of Choice we can construct the set of all its members step by step (as well as the set of all its subsets). The problematic nature of these assumptions, in particular the assertion of the existence of some property, and their use involving the *mathematical infinite*, represented an obstacle to the methodological reliability of the logicist programme.

The method of explicit definitions, i.e. the rejection of the impredicative ones, was meant to overcome the existential assumption by producing *constructions* corresponding

---

[12] We investigate recursive definitions more closely in Chapter 4.

to name-giving procedures for objects whose existence is already established. This illustrates the reduction of the existential claim to a definitional procedure. Hence, the Axiom of Reducibility represents the greatest difficulty, because it *presupposes* the possibility of definining a predicative function of interest. Carnap in his lecture in Königsberg attempted a consistent way to remove this axiom, using Ramsey's theory from *The Foundations of Mathematics*.[13] In order to remove the Axiom of Reducibility, Ramsey was in turn constrained to the assumption of existence of all properties, before they can be identified by definition. This means, again, to allow impredicative definitions. In rejecting this idea, Carnap maintains that this step means essentially to believe in a Platonic realm of ideas, whereas one should keep the Fregean requirement of accepting only what has been proven *in finitely many steps*. Carnap aimed at explaining how it is possible to give inductive general definitions without retaining conceptual absolutism, i.e. the idea of concepts existing before they can be constructed. In line with the general logicist approach, Carnap writes:

> [t]he verification of a universal logical or mathematical sentence does not consist in running through a series of individual cases, for impredicative definitions usually refer to infinite totalities. The belief that we must run through all the individual cases rests on a confusion of 'numerical' generality, which refers to objects already given, with 'specific' generality. We do not establish specific generality by running through individual cases but by logically deriving certain properties from certain others.[14]

In the formalist programme, the rejection of case-by-case analyses required reducing the complete verification of a statement about an arbitrary property to its *logical validity*, i.e. it being inferred by logical principles alone. The validity of a property defined impredicatively might be difficult or impossible only in individual cases, depending on the specific system in which one is working. Accordingly, formal theories require that definitions must be intended as *constructive* procedures and that those including totalities are valid as long as these can be guaranteed not to involve vicious circles or other paradoxical properties. Moreover, inside a system of deduction there is no primitive reference to the meaning of symbols (for this aspect, the programme has also passed in the literature with the name 'formalist').

2.4 Finitism

A second methodological approach to the problem of truth and correctness was proposed by the *finitist programme*: this was characterized by a stronger requirement on the notion of valid procedure in mathematics, requiring that everything involved in a mathematical definition needs to be reduced to *finite* properties, and everything should be proven by

---

[13] Ramsey (1926).
[14] (Carnap, 1931, p.51).

pure, *in principle mechanizable*, symbol manipulation techniques. This programme was started by David Hilbert on the assumption that the notion of infinity undermines the validity of deduction procedures and inferential methods.

According to Hilbert, to overcome the problems posed by paradoxes it was essential to guarantee an axiomatic formalization of theories in which only a completely clarified notion of infinity would be admitted. Under this stipulation, infinite objects and potentially infinite procedures were to be allowed only if reduced (or reducible in principle) to finite counterparts.[15] In particular, Hilbert refers to the use of infinite numerical series to define real numbers and the concept of real number itself, thought of as a completed totality. The introduction of the so-called method of *ideal elements* is Hilbert's way of clarifying the notion of infinity in cases like infinitely long lines and points at infinity from geometry; complex-imaginary magnitudes and ideal numbers from algebra; the entire analysis and Cantor's set theory:

> in analysis we deal with the infinitely large and the infinitely small only as limiting concepts, as something becoming, happening, i.e. with the potential infinite. But this is not the true infinite. We meet the true infinite when we regard the totality of numbers 1, 2, 3, 4, . . . itself as completed unity, or when we regard the points of an interval as a totality of things which exists all at once. This kind of infinity is known as actual infinity.[16]

Errors arise in the misapplication of the infinite: in particular, material logical deduction produces errors in the form of arbitrary abstract definitions involving infinitely many objects. For the relation of logical deduction to be valid, properties of the extra-logical objects to which logical signs are applied need to be clearly established. For the rest, logic (and mathematics with it) is a pure, completely determined game of signs, with finitary and directly constructive procedures:

> [a]lthough the content of a classical mathematical sentence cannot always (i.e. generally) be finitely verified, the formal way in which we arrive at the sentence can be. Consequently, if we wish to prove the validity of classical mathematics, which is possible in principle only by reducing it to the a priori valid finitistic system [...], then we should investigate, not statements, but methods of proof.[17]

As an effect of this paradigmatic shift from the logicist inspired unique system of thought to the finitist system of manipulable symbols, logic becomes a way of axiomatizing independent theories, with valid inferential structures built on finitistic, arithmetic-combinatorial bases. To this aim, the Hilbertian method can be reduced to the following basic steps:

- the formulation of a logical vocabulary;
- the unambiguous characterization of a combination of symbols (meaningful formulas);

---

[15] Hilbert (1983).
[16] (Hilbert, 1983, p.188).
[17] (von Neumann, 1983, p.62).

- the description of a construction procedure called 'proving' allowing to formulate all the valid formulas;
- the acceptance of mathematical statements only on the basis finitary proving methods.

Given this formulation of a logical method for the sciences, Hilbert illustrated his programme at the Congress of Mathematics held in 1928 in Bologna, Italy, with the aim of proving the following results:

Thesis 3 (Completeness) *All true mathematical statements can be proven.*

Thesis 4 (Consistency) *Only true mathematical statements can be proven.*

Thesis 5 (Decidability) *A decision procedure exists to decide the truth or falsity of any given mathematical proposition.*

The birth of logic as meta-theoretical analysis of scientific theories, i.e. the ability to prove completeness and consistency, is based on the last requirement. The effective check provided by decidability, together with consistency, will ensure that no formula derivable in the system can ever be an equation of the form $1 = 0$. Such a proof must of course be characterized by those finitary terms that the method requires:

> The real problem is then that of finding a finitary combinatorial proof of consistency.[18]

It is on the identification of the essential conditions for such a procedure that intuitionism will proceed.

2.5 Intuitionism

The third approach to determine truth and correctness in mathematics would focus in particular on the need to redefine the former notion on a procedural basis, with the aim of maintaining control of the definitional process. Brouwer's philosophy is at the basis of this idea of truth defined by assertion conditions on contents. A formal semantic translation was offered by Heyting's interpretation of intuitionistic logic from around 1930. Kolmogorov further interpreted it in terms of problems and their solutions. The so-called Brouwer-Heyting-Kolmogorov semantics defines the truth of a proposition as the existence of a proof-object for it.[19] According to the intuitionistic perspective, to reduce a definition to a well-defined and complete procedure might not be sufficient if its result does not reduce to determining the conditions under which contents can be known.

---

[18] (von Neumann, 1983, p.64).

[19] See Brouwer (1925a,b) for the theoretical foundation of intuitionism, Heyting (1930) for its formal translation, and Kolmogorov (1932) for the problem-task interpretation. Details on the BHK formal semantics of proofs are provided in Chapter 7, but for now we are only interested in presenting the guiding principles of intuitionism to the problem of determining true sentences in mathematics.

To illustrate this point, Heyting in his lecture considers the definition of a real number in terms of assigning to every rational number either the predicate LEFT or RIGHT, in order to preserve the natural order of rationals. Its reduction to the procedure enclosing the Euler's constant C within an arbitrarily small rational interval can be obtained by computing an always smaller series of rational intervals. The procedure is still insufficient to decide for an arbitrary rational number A whether it lies left or right of C or is perhaps equal to C.[20] The reason for this uncertainty is the generalized application of the Aristotelian law of excluded middle, i.e. that any proposition either it is true or it is false, formally $A \vee \neg A$. In its application for the calculation of C, the rejection of this law means it is impossible to affirm that either a number n of computation steps proves that $A < C$ or $A > C$, or that such n does not exist until that is actually shown to be the case, and hence the contradictory case can be excluded. The clue to reject the famous Aristotelian principle is given again in terms of the notion of infinity:

> We can drop the requirement that the series of predicates be determined to infinity by a rule. It suffices if the series is determined step by step in some way, e.g. by free choices. I call such sequences 'infinitely proceeding'.[21]

This description of infinite objects such as choice sequences, i.e. in terms of step-by-step procedures, has an effect on their definition. An infinite object cannot be regarded as the collection of its members, which is a meaningless statement if such object is not considered as existing in itself. Instead, an object is defined by the construction on the basis of previously defined elements, down to elementary ones:

> Impredicative definitions are made impossible by the fact, which intuitionists consider self-evident, that only previously defined objects may occur as members of a species [set].[22]

To admit no infinitely proceeding sequence in mathematics means, in other words, to admit only rule-determined sequences, to which a number belongs if and only if there is a rule which allows one to actually determine all predicates of the sequence successively. When applied to some specific case like 'whether or not the sequence 0123456789 occurs in the decimal expansion of π'[23] this requirement corresponds to the formulation of a binary YES/NO problem and the task of finding a solution (proof) to it. A problem is formulated by an intention (expectation) to find its fulfilment (solution): the solution is obtained if a construction is provided, or else it must be proved that the intention leads to a contradiction. This amounts to a reduction of solvability to provability. Truth, as the property of propositions providing solutions to problems, is thus in turn reduced to proofs:

[20] (Heyting, 1983, p.54).
[21] (Heyting, 1983, p.55).
[22] (Heyting, 1983, p.57).
[23] (Heyting, 1983, p.58).

Definition 16 (Truth as Proof) *Proposition A is true if and only if there exists a proof a of A.*

The constructive interpretation of the infinite by means of so-called *lawlike sequences* reflects this concept of the truth of a property being reduced to a construction answering its validity:

> [a lawlike sequence] might be described as a sequence [a mapping associating with every natural number a mathematical object belonging to a certain well-defined set] which is completely fixed in advance by a law, i.e. a prescription (algorithm) which tells us how to find for any n member of $\mathbb{N}$ the nth member of the sequence.[24]

When this problem is generalized to theories, Hilbert's aim of determining the truth of any proposition A within a given set of axioms Γ corresponds to constructing a finite proof procedure to check whether A is a theorem within the theory described by Γ. In view of the restriction required by the intuitionistic foundation, several theorems and axioms of classical mathematics could not be proven valid, because not 'constructible', as for example the Law of Excluded Middle and the Fixed Point Theorem of Topology. In view of the idea of providing a 'foundation' (rather than a method or instrument of proceeding in the construction of science), the intuitionistic programme failed in that all classical mathematics cannot be 'constructed'. On the other hand, the failure of formalism was determined by the impossibility of reducing all mathematics to logical sentences. And the failure of formalism was dictated by Gödel's First Incompleteness Theorem.[25]

The foundational crisis of mathematics amounted to the quest for establishing all valid formulas of theories, avoiding the various traps of paradoxes, circular definitions, and the infinity. In particular, Hilbert was looking for a *mechanical way* to determine the provability of every possible mathematical sentence of any given system. The formalization of this task requires set-theoretical tools which we overview in Chapter 3.

Exercises

Exercise 1 *Explain informally how consequence and inference differ in explaining mathematical truths.*

Exercise 2 *Explain informally what the soundness and completeness problems for mathematical theories are.*

Exercise 3 *Illustrate the main principles of the logicist programme in the foundations of mathematics.*

---

[24] (Troelstra, 1969, p.17).
[25] Gödel (1931).

Exercise 4 *Assuming the definition of the cardinality of a set as the number of its elements, and Frege's definition of number as the extension of a concept, which concepts are expressed by the cardinality of an empty set and of an infinite set?*

Exercise 5 *Consider the following sentence: 'This sentence is false.' Explain how this generates a paradox.*

Exercise 6 *Illustrate the main principles of the finitist programme in the foundations of mathematics.*

Exercise 7 *Explain informally what decidability for sentences of mathematical theories means.*

Exercise 8 *Consider the following definition of the set S: the set of all sets that are greater than S. What type of definition is this and what problems can arise in connection to it?*

Exercise 9 *Illustrate the main principles of the intuitionist programme in the foundations of mathematics.*

Exercise 10 *Why is a mechanical procedure essential according to Hilbert for deciding truth?*

3 Computing and Deciding

Summary

This chapter illustrates the basic tools of computability theory, essential to the formulation of the decision problem and the definition of the notion of computable function.

3.1 Enumerability

The limits to validity identified during the foundational crisis were mainly due to the role that infinity played in the definition of mathematical concepts. This notion of infinity has been crucial in mathematics since antiquity, and it had been the source of paradoxes at least since Zeno. Its presence would return crucially in the seventeenth century with the development of calculus by Newton and Leibniz, and later in the theory of sets by Georg Cantor (1845–1918). His famous theorem based on the diagonalization method is a milestone for mathematics and it allows us to review some essential techniques in the formulation of limiting results related to computability. The method involves essentially the use of the infinite by referring to the enumeration of an infinite list of objects. Such enumeration is expressed in terms of a well-defined function which returns the intended set as result. As required by Hilbert, this notion of infinite list is then reduced to an effectively computable method if it can be guaranteed that each entry of the list comes at a certain point after a finite number of members of the list has been enumerated after the first one. Cantor's diagonal method shows that even in this way it is not possible to arrange all sets of positive integers in a single infinite list.

Let us start by some fundamental notions of naive set theory. Recall that a set is formally intended as a collection of objects, known as the elements or members of the set. We denote the membership of an object s in the set S by the formula $s \in S$. A set is

On the Foundations of Computing. Giuseppe Primiero, Oxford University Press (2020). © Giuseppe Primiero.
DOI: 10.1093/oso/9780198835646.001.0001

well defined if and only if, for any given object s, we can say for certain object whether or not it is a member of the set, i.e. whether $s \in S$ or $s \notin S$. This means to be able to list all elements of S, or enumerate them.

Definition 17 (Enumerable Set) *An enumerable or countable set S is one whose members can be enumerated or arranged in a list.*

The list has to include all and only the elements of the set, in a way such that after some finite number of steps, one can reach any of the elements in that set. The list is unordered, i.e. its members do not have a fixed position in the list, and copies of members do not count. An example is the set of natural numbers smaller than 10, or that of numbers which can divide 2 without returning a rational. A set with an infinite number of elements can also be arranged as a list, e.g. the set of natural numbers $\mathbb{N}$:

Definition 18 (Denumerable Set) *A denumerable or enumerably infinite set is one with an infinite number of members arranged in a list.*

Sets can be enumerated by functions as well:[1]

Definition 19 (Function) *A function f is an assignment of values to arguments. The set of all valid arguments of a function is the domain D of the function f. The set of all the values the function assigns to D is the range R of the function f.*

Definition 20 (Total Function) *A total function f(d) from the domain D to the range R is defined for every element $d \in D$.*

The enumeration of a set by a function needs to be complete: every element in the set needs to appear at least once. By the definition of set, a redundant list of elements obtained by a well-defined function is still an enumeration of the elements in that set. On the other hand, there are functions which are not defined over every possible element:

Definition 21 (Partial Function) *A partial function f(d) from the domain D to the range R is defined for some element $d \in D$.*

Definition 22 (Surjective or Onto Function) *A surjective function f(d) from the domain D to the range R has at least one element $d \in D$ assigned for every element $r \in R$.*

Definition 23 (Injective or 1–1 Function) *An injective function f(d) from the domain D to the range R has at most one element $d \in D$ for every element $r \in R$.*

Definition 24 (Inverse Function) *The inverse function $f^{-1}(d)$ from the domain D to the range R is defined by letting $f^{-1}(r)$ be the one and only d such that $f(d) = r$, if such an element d exists.*

---

[1] In the following we use the terms *domain* and *range* to denote respectively the set of possible input and output values of functions; totality (or partiality) of a function is determined by the domain being well defined. Another terminology distinguishes the *source* and the *target* as input and output sets, while *domain* and *range* are strictly used to refer to subsets of source and target containing respectively valid input and actual outputs.

In the following, we constrain ourselves to functions whose domain is the set of positive integers:

Definition 25 (Function of Positive Integers) *A function f of positive integers $\mathbb{Z}$ encodes an arrangement of the members of its range R.*

Let us consider some examples:

- the set E of even positive integers is encoded by the function $f(n) = 2(n)$;
- the set O of odd positive integers is encoded by the function $f(n) = (2n - 1)$;
- the whole set $\mathbb{Z}$ of positive integers is encoded by the function $f(n) = n$ (i.e. the identity function);
- the whole set $\mathbb{Z}$ of positive integers is (also) encoded by the function

$$g(n) = \begin{cases} n + 1, & \text{if } n \text{ is odd} \\ n - 1, & \text{if } n \text{ is even} \end{cases}$$

Every function can be defined as partial. Consider the following examples:

- the set E of even positive integers is encoded by the partial function $j(n)$

$$j(n) = \begin{cases} n, & \text{if } n \text{ is even} \\ \text{undefined otherwise} \end{cases}$$

- any subset S of the set $\mathbb{Z}$ positive integers is encoded by the partial function $k(n)$

$$k(n) = \begin{cases} n, & \text{if } n \text{ is in the set } S \\ \text{undefined otherwise} \end{cases}$$

Every subset of the set of positive integers is enumerable, i.e. a function can be defined for every subset such that the function gives that set as its range. Hence, the following definition:

Definition 26 (Enumerable Set by a Function) *A set S is enumerable if and only if for every $s \in S$ there is at least one positive integer $n \in \mathbb{Z}$ such that $f(n) = s$.*

In view of the definition of an enumerable set as the ordered list that can be obtained as the result of applying a function to integers, one also has the following:

Definition 27 (Enumerable Set) *A set S is enumerable if and only if S is the range of some function f of positive integers, i.e. $f(n) = s$ for every $s \in S$.*

Such function f defining the elements in S is called the characteristic function of S:

Definition 28 (Characteristic Function) *A function f is called the characteristic function of a set S if and only if*

$$f(s) = \begin{cases} 1, & \text{if } s \in S \\ 0, & \text{if } s \notin S \end{cases}$$

3.2 Encoding

The notion of function is generalized by defining it for more than one argument. A function $f(m,n)$ can be conveniently defined as a one-place function with ordered pair of integers as argument (and so for many arguments). The inverse generalization is to look at a function of one argument as the code that yields any given pair (or set) of values as its range. In this case we call the argument of that function the code for the given pair:

Definition 29 (Code Number) *Given a function f which enumerates pairs of positive integers (m,n), any argument s such that $f(s) = (m,n)$ may be called the code number for the pair (m,n). Applying the function f means to decode s; the other way round means to encode the pair (m,n).*

In order to define a function that takes a pair of integers as argument and yields an integer as value, one can encode the set of ordered pairs of positive integers. A possible method to do this consists in listing all the pairs of positive integers according to the following rule:[2]

1. enter all the pairs the sum of whose entries is 2;
2. enter the pairs the sum of whose entries is 3 and so on;
3. for entries with equal sum, enter first the pair with lower first entry:

$$(1,1),(1,2),(2,1),(1,3),(2,2),(3,1),(1,4)\ldots$$

Then a code for (m,n) is the number s corresponding to the position in which the pair (m,n) is located. That number is obtained by counting all the pairs $m+n-2$ whose sum is $m+n-1$. For example, for the pair $(m=1,n=1)$ there are $1+1=2-2=0$ pairs to go through whose sum is $1+1-1=1$. The general formula is given as

$$[1+2+\ldots(m+n-2)]+m$$

The code J for any item in this list is as follows:

$$J(m,n) = (m^2 + 2mn + n^2 - m - 3n + 2)/2$$

A generalization of the previous case is to encode the set of finite sequences of positive integers. Take $G_1(n)$ to be the 1-term sequence n. G_2 is the function enumerating all the 2-tuples or pairs from Cantor's zig-zag list:

$$(1,1),(1,2),(2,1),(1,3),(2,2),(3,1),(1,4)\ldots$$

---

[2] This method is illustrated in (Boolos et al., 2002, chapter 1).

G_3 is the function enumerating the triples obtained by substituting in the previous list the second entry n with the nth component in the same list:

$$(1,(1,1)),(1,(1,2)),(2,(1,1)),(1,(2,1)),(2,(1,2)),(3,(1,1)),(1,(1,3))\ldots$$

and so on for every G_i. An encoding of all these sequences by a pair of positive integers can be obtained by encoding any sequence G of length k by the pair (k,a) where $G_k(a) = s$. In other words: start by the original listing of pairs and replace any pair (k,a) by the ath item on the list of k-tuples:

- replace $(1,1)$ by the first item 1 in the list of 1-tuples;
- replace $(1,2)$ by the second item 2 in the list of 1-tuples;
- replace $(2,1)$ with the first item $(1,1)$ in the list of 2-tuples:

$$(1),(2),(1,1),(3),(1,2),(1,1,1),(4)\ldots$$

3.3 Diagonalization

Given the definition of enumerable set, it is easy to understand the notion of a nonenumerable set, i.e. a set for which no function can be given whose range contains all the elements in that set. The existence of such sets is exemplified by the following theorem:

Theorem 1 (Cantor's Theorem) *The set P^* of all sets of positive integers is not enumerable.*

Proof. The proof is by contradiction. Build the infinite list L of sets of positive integers $S_1, S_2, S_3, \ldots$. This will contain all the possible subsets of the set of integers (odd, even, and so on). Then define the set $\Delta(L)$ as follows:

Definition 30 (Diagonal List) *For each positive integer n, n is in the diagonal list $\Delta(L)$ if and only if n is not in S_n.*

Let us proceed by constructing $\Delta(L)$. Let us assume that S_1 is the list of prime positive integers $S_1 = \{2, 3, 5, \ldots\}$; then $1 \in \Delta(L)$ because $1 \notin S_1$. Take now S_2 and assume this is the list of even positive integers $S_2 = \{2, 4, 6, \ldots\}$; then $2 \notin \Delta(L)$ because $2 \in S_2$. Take now S_3 and assume this is the list of odd positive integers $S_3 = \{1, 3, 5, \ldots\}$, then $3 \notin \Delta(L)$ because $3 \in S_3$. And so on. Then, provided L is the list of *all* sets of integers, $\Delta(L) \in L$ i.e. $\exists m.S_m = \Delta(L)$ for some positive integer m. But for the definition of $\Delta(L)$

Definition 31 $m \in \Delta(L)$ *if and only if* $m \notin S_m$.

This contradicts the following:

Definition 32 $m \in \Delta(L)$ *if and only if* $m \in S_m$.

Hence if $\Delta(L) \in L$ we have a contradiction, therefore $\Delta(L)$ cannot be in L. This demonstrates that the set L of all sets of positive integers is not enumerable. □

This proof can be illustrated applying the diagonalization method. Let the list of sets of positive integers

$$L = S_1, S_2, S_3, \ldots$$

be represented by functions $s_1, s_2, s_3, \ldots$ ranging over $\{0, 1\}$ according to the following definition:

$$s_n(p) = \begin{cases} 1, & \text{if } p \text{ is in } S_n \\ 0, & \text{if } p \text{ is not in } S_n \end{cases}$$

As an example, for S_3 the list of odd positive integers $S_3 = \{1, 3, 5, \ldots\}$, the function s_3 will give the following values:

$$s_3(1) = 1; s_3(2) = 0; s_3(3) = 1; \ldots$$

Order now vertically the list of functions $s_1, s_2, s_3, \ldots$ and let the list

$$s_n(1), s_n(2), s_n(3), \ldots$$

be the nth row in that list. Then this row will be a list of $0, 1$, according to the presence of each entry p at $s_n(p)$ in the set s_n. The *diagonal list* is obtained by taking the following elements:

$$s_1(1), s_2(2), s_3(3), \ldots s_n(n).$$

i.e. the first entry in the first list, the second entry in the second list, up to the nth entry in the nth list. Then the *anti-diagonal list* is obtained by inverting the values of each entry: change to zero the values one and to one the values zero of the items in the diagonal list. This can be obtained as follows:

$$s_m(1) = 1 - s_1(1),$$
$$s_m(2) = 1 - s_2(2),$$
$$s_m(3) = 1 - s_3(3),$$

$$\vdots$$

where each $s_n(n)$ is a zero or a one. The sequence defined as the antidiagonal list does not appear in any place of the list L. If it should, at row m we would have

$$s_m(1) = 1 - s_1(1),$$
$$s_m(2) = 1 - s_2(2),$$
$$s_m(3) = 1 - s_3(3),$$

$$\vdots$$

$$s_m(m) = 1 - s_m(m)$$

The mth item in this list S_m gives the identity $1 = 0$, because if $s_m(m) = 1$ then $1 - s_m(m) = 0$ and if if $s_m(m) = 0$ then $1 - s_m(m) = 1$.

The argument at the basis of Theorem 1 can be offered now in a short version as follows:

Proof. For every set S the power set $P(S)$ of S, i.e. the set of all subsets of S is larger than S itself. Let f be a function $f : S \mapsto P(S)$, then it is not the case that for every element $p \in P(S)$ there is an element in $s \in S$ such that $f(s) = p$, i.e. some subset of S is not in the image of the function. Let us consider the set

$$T = \{ s \in S \mid s \notin f(s) \}.$$

T is not in the image of f: for all $s \in S$, either s is in T or it is not. In both cases, $f(s) \neq T$, because: if $s \in T$, then by definition of T, $s \notin f(s)$, so $T \neq f(s)$ since $s \in T$ but $s \notin f(s)$; if $s \notin T$, then by definition of T, $s \in f(s)$, so $T \neq f(s)$, since $s \notin T$ but $s \in f(s)$. $\square$

3.4 The Decision Problem

The result of Cantor's Theorem leads to the idea that there are sets whose elements cannot be enumerated. Provided enumerability allows one to define a characteristic function for the relevant set, nonenumerability means nondefinability. Consider the set S of all sets: according to the previous argument, then its power set $P(S)$ is the set of all subsets of all sets. As S is the set of all sets, it is supposed to include $P(S)$ to be complete; but $P(S)$ should also be bigger than S. This shows informally that the notion of 'set of all sets' is inconsistent. This is what Russell's paradox has shown: if any definable subclass of a set is a set, contradictions arise. This last claim can be shown by the use of the following principle:

Definition 33 (Comprehension Principle) *Given any set S and predicate ϕ, there is a set S' which is a subset of S such that x is a member of S' if and only if $x \in S$ and $\phi(x)$ holds. By the axiom of extensionality this set is unique.*

The unrestricted version of the Comprehension Principle let impredicativity arise, because it says that every subset defined by a predicate is itself a set:

Definition 34 (Unrestricted Comprehension Principle) *Given any predicate ϕ, there is a set S such that x is a member of S if and only if $\phi(x)$ holds. By the axiom of extensionality this set is unique.*

The basic answer to the problem of impredicativity and paradoxes obtained by the diagonalization method is represented by the notion of *effective computability*. Accordingly, the problem of enumerability and of definability, reduces to the problem of establishing whether the characteristic function for any given set can be in some sense qualified as effectively computable. The computability of the characteristic function for the set of theorems and (assuming soundness and completeness) of the consequence set of any given theory is the precise formulation of Hilbert's problem of decidability (see Thesis 5):

Proposition 2 (Decision Problem) *Can an effective method be formulated that applied to any finite set of sentences* Γ *and a sentence* ϕ *within a system of logic F would establish in a finite amount of time whether* $\Gamma \vDash_F \phi$?

For example: one wants to decide given a formula ϕ of number theory whether it holds for any number x, i.e. if it is valid.

One can also understand the question in purely mechanical terms: imagine the system of reference as a set of axioms and rules denoted by Γ; a system of axioms and rules is logically equivalent to a 'machine'; and take *any* output ϕ. We want to check, if any such ϕ is or is not an output of our 'machine' Γ. Hence, one is looking for a method that can always establish if something is or is not a valid output of a given program for any machine. This gives us the mechanical version of the decision problem:

Proposition 3 (Mechanical Decision Problem) *Could there be a generalized 'calculation procedure' that would tell us whether any output* ϕ *can be obtained by any machine* Γ *with a given input* ι?

The notion of effective method or calculation procedure at the basis of the decision problem needed clarification. In other words, the following question emerged: what does it mean that a function is effectively calculable? Our next task is to formulate a mathematically precise answer to this question.

Exercises

Exercise 11 *Give at least one example of an enumerable set and one of a denumerable set with elements in* $\mathbb{N}$.

Exercise 12 *Give at least one example of a total function and one of a partial function with domain in* $\mathbb{Z}$.

Exercise 13 *Consider the following function with domain in* $\mathbb{N}$:

$$f(n) = \begin{cases} 1, & \text{if } n = 1 \\ n - 1, & \text{if } n > 1 \end{cases}$$

Qualify this function as either injective or surjective and explain why.

Exercise 14 *Give at least one example of an injective function over* $\mathbb{Z}$.

Exercise 15 *Give at least one example of a function over* $\mathbb{Z}$ *and define its inverse.*

Exercise 16 *Enumerate the set of the successors of the odd numbers by a function.*

Exercise 17 *Define a subset of* $\mathbb{Z}$ *by its characteristic function.*

Exercise 18 *Show formally how the characteristic function of the diagonal list generates an arithmetical contradiction.*

Exercise 19 *What does it mean for a set to be nonenumerable?*

Exercise 20 *Explain the decision problem.*

4 What is Computable?

Summary

In the previous chapter we investigated the formal origins of the decision problem and anticipated its *mechanical* version. In this chapter, we define formally the notion of *computable function* through its inductive and recursive definitions. We cover the construction schemas to define total and partial computable functions. We conclude by explaining how recursive definitions are general and equivalent to other formulations, a result known as *Church's Thesis*.

4.1 Mathematical Induction

Our historical and conceptual analysis so far has illustrated the rising need among mathematicians and logicians for the determination of procedures considered admissible in the sciences, and in arithmetic in particular. Examples of crucial milestones in this context are impredicative definitions and the role of the infinite in logical derivations. The aim was the identification of valid *definitional procedures* and *steps of computation* from the *admissible primitive notions*, defining in turn new admissible terms. In the case of definitions this was implemented by the notion of non-vicious predications: any term is required to be defined from building blocks not including any totality to which the term belongs. Computational procedures, on the other hand, refer to steps to be performed in terms of definite and explicit instructions to compute new values. Processes in terms of computational steps that can be *effectively* formulated reflect an intuitive notion:

Definition 35 (Effectively Computable Function) *A function f with domain and range in $\mathbb{Z}$ is called effectively computable if a list of definite and explicit instructions can be given to compute the value of $f(n)$, for any possible argument n.*

On the Foundations of Computing. Giuseppe Primiero, Oxford University Press (2020). © Giuseppe Primiero.
DOI: 10.1093/oso/9780198835646.001.0001

Validity for any possible argument n of the function means simply that it needs to be completely specified what the function does for any argument given as input: at each step and for each possible argument, it can be spelled out in detail the procedure to obtain the value of $f(n)$, in such a way that everything needed for such a computation is given by the definition of f itself. External problems, such as the fact that the computation (or the computator) will run out of time or energy or memory are irrelevant to the definition of the procedure. In a word, at every step of the computation the process is finite and completely defined. A second important condition is that the *representation system*, or *encoding*[1] used for the input n (or (n, m) in case of a pair of arguments) to which the function f is applied is entirely irrelevant to the computation itself. One can choose any possible representation and then show that it can be translated into other forms; obviously, the more general such representation, the better.

The formal tool identified during the 1930s to express effective calculability was *recursion* and it was already available to mathematicians since Euclid's *Elements* in the form of definition by *induction*. Early attempts at using this method to provide solid foundations for the theory of numbers were due to Weierstrass: in his lectures delivered in Berlin, he defined irrational numbers inductively from the rationals, and these in turn from the integers.[2] Also Cantor and Dedekind approached the problem of irrational numbers by using the integers as their inductive base.[3] In the second half of the twentieth century, the method of induction was also used by Grassmann, Peirce, and Peano:[4] Peirce and Grassmann used recursion to define addition and multiplication and used mathematical induction to prove the laws of arithmetic; Dedekind identified the axioms for zero 0 and the successor function S. Establishing the first element and the iterative structure to generate every other element in the set $\mathbb{N}$ of natural numbers, Dedekind designed them as a unique infinite sequence without cycles. Following Grassmann, he then used recursion to define addition, multiplication, and exponentiation using the successor function.[5] Peano, in his axiomatic foundation of arithmetic, provided a similar construction starting from the number one, the successor operator and equality. Among his axioms, he uses induction and recursion to define multiplication based on addition and exponentiation based on multiplication.[6]

Let us provide a general axiomatic definition of the natural numbers:

Definition 36 *The primitive axioms for the arithmetical hierarchy are:*

1. *0 is a number*
2. *If x is a number, then $S(x)$ is a number*
3. *$0 \neq S(x)$, for every x*

---

[1] See Chapter 3.2.
[2] See Tweddle (2011).
[3] Cantor (1872); Dedekind (1960, 1888).
[4] Grassmann (1861); Peirce (1881); Peano (1889).
[5] Grassmann (1861).
[6] See Kennedy (1973).

4. *If $S(x) = S(y)$ then $x = y$*

5. *If $x \neq 0$ then there is a y such that $x = S(y)$*

6. *If a set $\mathbb{N}$ contains 0, and $S(x)$ for every x, then every number is in $\mathbb{N}$.*

Let us provide here a basic formal definition of recursion as induction:

Definition 37 (Recursion as Induction) *A function on numbers $f(n)$ is recursively defined if and only if*

1. *an explicit definition for $f(0)$ is given, and*

2. *$f(n+1)$ is defined in terms of $f(n)$ by means of already defined functions.*

Example 1 (Addition by Recursion on Successor) *Assume one already knows what the successor function is, e.g. the function denoted as n' which allows it to move one step up in an ordered sequence of natural numbers when moving from n; then addition can be defined recursively in terms of successor, as follows:*

$$m + 1 = m'$$
$$m + n' = (m + n)'$$

A concrete instance of this schema for the definition of the number 2 is as follows:

$$0 + 1 = 0' = 1$$
$$0 + 1' = (0 + 1)' = 2$$

Defining the sum function by induction on the successor function allows any of the other basic arithmetical functions to be defined.

Example 2 (Multiplication by Recursion on Addition) *Assuming addition has been defined, multiplication is defined as follows:*

$$m \times 0 = 0$$
$$m \times n' = (m + (m \times n))$$

A concrete instance of this schema for the definition of the number 3 is as follows:

$$1 \times 0 = 0$$
$$1 \times 0' = (1 + (1 \times 0)) = 1$$
$$1 \times 1' = (1 + (1 \times 1)) = 2$$
$$1 \times 2' = (1 + (1 \times 2)) = 3$$

Example 3 (Exponentiation by Recursion on Multiplication) *Assuming multiplication has been defined, exponentiation is defined as follows:*

$$m^0 = 1$$
$$m^{(n')} = (m \times m^{(n)})$$

A concrete instance of this schema for the definition of the number 4 is as follows:

$$2^0 = 1$$
$$2^{0'} = (2 \times 2^{(0)}) = 2$$
$$2^{1'} = (2 \times 2^{(1)}) = 4$$

Since the seminal works by Grassmann and Dedekind, and up to the contributions of Skolem, Gödel, and Kleene,[7] the definition of computability by induction has been more precisely specified in terms of *primitive recursion*, according to the following steps:

1. introduce the so-called *basic recursive functions*;
2. define construction schemas for the *total effectively computable functions*, also known as *primitive recursive functions*;
3. define construction schemas for the *partial effectively computable functions*, simply known as *recursive functions*.

We explore these steps in the following sections of this chapter.

4.2 Primitive Recursion

Let us start with the first step:

Definition 38 (Basic Functions) *The zero Z, successor S, identity I, and projection functions π_i^n are called basic functions. They can be computed in one step.*

These are primitive, in the sense that their definition is immediate:

- for the Z function, given any input $n \in \mathbb{N}$, the function returns 0;
- for the S function, given any input $n \in \mathbb{N}$, the function returns n', i.e. the next element in the set;
- for the I function, given any input $n \in \mathbb{N}$, the function returns n;
- for the π_i^n function, with $n = \{1, \ldots, n\}$, the function returns the ith element in the set of n elements.

From the basic functions we can build new effectively computable functions by using construction schemas. The first operation which can be performed on basic functions to obtain more complex functions is *composition* (or *substitution*).

Definition 39 (Composition) *If f is a function of m arguments and each $g_1, \ldots, g_m$ is a function of n arguments, then a function obtained by composition from $f, g_1, \ldots, g_m$ is the function h defined as follows:*

---

[7] Skolem (1923); Gödel (1931); Kleene (1936a,b).

$$h(x_1,\ldots,x_n) = f(g_1(x_1,\ldots,x_n),\ldots,g_m(x_1,\ldots,x_n))$$

If functions g_i are all effectively computable and f is effectively computable, then h is effectively computable.

Composition of two functions f and g is also denoted by $f \circ g$. Taken any function of m arguments, if each such argument is obtained by a computable function and the function in question is itself computable, applying the latter to its arguments returns a computable function.

Example 4 *Given $f(x) = 2x + 3$ and $g(x) = -x^2 + 5$, $h = (f \circ g)(x)$ is defined as follows:*

$$h = (f \circ g)(x) = f(g(x))$$
$$= f(-x^2 + 5)$$
$$= 2(-x^2 + 5) + 3$$
$$= -2x^2 + 10 + 3$$
$$= -2x^2 + 13$$

Other effectively computable functions can be obtained by recursive equations.

Definition 40 (Recursion) *The process of defining new functions from old ones called primitive recursion is defined by the following equations:*

$$h(x,0) = f(x)$$
$$h(x,y\prime) = g(x,y,h(x,y))$$

The function h is said definable by primitive recursion from functions f and g.

Taken any computable function f on input x which defines the value of a function h on input 0, the value of h on the successor value is given recursively as a function g on the input x and the predecessor of the current value for h. As we know, the successor function is computable, and assuming a function is defined on 0, it is thereby defined on every number.

Example 5 (Addition by Recursion)

$$f(x,y) = x + y$$
$$f(x,0) = x$$
$$f(x,y+1) = (f(x,y)) + 1$$

Definition 41 (Primitive Recursive Functions, Dedekind (1888); Skolem (1923); Gödel (1931)) *The class of primitive recursive functions is the smallest class including the basic functions and closed under composition and recursion.*

All primitive recursive functions are effectively computable. The basic functions plus the composition and recursion schemas give us the set of *all effectively computable total functions*. This class will include summation, product, power, predecessor, and difference.

4.3 Partial Recursion

The previous schemas defining primitive recursion produce total functions from total (basic) functions. A new schema can be added which, applied to an effectively computable function, gives as result either a total or a partial function, depending on the input:

Definition 42 (Minimization) *Given a function f of n arguments, minimization yields a total or partial function of $n + 1$ arguments as follows:*

$$\mu[f(x_1,\ldots,x_n)] = \begin{cases} y, & \text{if } f(x_1,\ldots,x_n,y) = 0 \text{ and for all } t < y \\ & f(x_1,\ldots,x_n,t) \text{ is defined and } \neq 0 \\ \text{undefined} & \text{if there is no such } y \end{cases}$$

If f is an effectively computable total or partial function, then $\mu[f]$ will be a computable, respectively total or partial, function.

The minimization schema works as follows: given arguments $(x_1,\ldots,x_n)$, the function searches the first additional argument y such that it gives output 0; when $(x_1,\ldots,x_n)$ is not in the domain of $\mu[f]$, there will be no such y. This can be due to one of the following reasons:

1. for any additional argument, the function is defined but never equal to 0;
2. or for some additional argument i, the functions up to

$$f(x_1,\ldots,x_n,i-1)$$

are defined and different than 0, but

$$f(x_1,\ldots,x_n,i)$$

is undefined.

In both cases the computation process will keep searching for the next value $i + m$ such that the first condition in the definition schema is satisfied.

Example 6 (Addition is not preserved by Minimization) *Consider minimization applied on the addition function*

$$\mu[+(x_1,\ldots,x_n)]$$

where all x_n are natural numbers. Then consider the result of the schema, which searches for some $y \in \mathbb{N}$ such that $+(x_1,\ldots,x_n,y) = 0$. This can be the case if and only if $x_1 = \cdots = x_n = y = 0$. Hence, it is not the case that minimization is well defined over every input of addition.

Example 7 (Multiplication is preserved by Minimization) *Consider minimization applied on the multiplication function*

$$\mu[\times(x_1,\ldots,x_n)]$$

Multiplication is total, because for every x, it is enough to include $y = 0$ to have $x \times 0 = 0$.

The last example shows how minimization works for total functions:

Definition 43 (Minimization for Total Functions) *If the function f is total, the definition of the minimization schema reduces to the following:*

$$\mu[f(x_1,\ldots,x_n)] = \begin{cases} \text{the smallest } y \text{ such that } f(x_1,\ldots,x_n,y) = 0 \text{ if it exists} \\ \text{undefined otherwise} \end{cases}$$

The total function f is called regular if for every $x_1,\ldots,x_n$ there is a y such that $f(x_1,\ldots,x_n,y) = 0$; if f is a regular function then $\mu[f]$ is total.

Adding the minimization schema, one includes partial functions that are effectively computable:

Definition 44 (Kleene Normal Form, Kleene (1936a)) *The class of recursive functions is the smallest class including the basic functions (zero, successor, projection) and closed under composition, primitive recursion and minimization.*

Note that given a set S whose members are all the elements characterized by a given predicate P, we can associate the definability of S with recursiveness of P:

Definition 45 (Recursive Predicate) *A predicate P is recursive if and only if its characteristic function is recursive.*

In other words, if it is possible to define a function f such that it determines for every element s whether $s \in S$, where S is the set defined by P, and f is recursive, then P is recursive as well.

The notion of (deterministic) function can be defined as a relation between domain and range where for every element in the domain there is at most one element in the range for which that function holds. The generalization to the notion of relation is obtained through the recursiveness of the relevant predicate:

Definition 46 (Decidable Relation) *A relation R is effectively decidable if and only if its defining predicate is recursive.*

With the minimization schema and the recursive definition of possibly partial functions, we have the weakening of the previous property:

$$(\lambda x. M[x]) \rightarrow (\lambda y. M[y])$$

α-conversion: to rename bound variables in λ-terms.

$$((\lambda x. M)N) \rightarrow (M[x := N])$$

β-reduction: to apply a function $\lambda x. M$ to a term N.

Figure 4.1 Rules of Church's Simple Lambda Calculus

Definition 47 (Semi-Decidable Relation) *A relation R is effectively semi-decidable if it is obtained from an effectively decidable relation by unbounded existential quantification: in particular, the relation $R(x, t)$ that produces in t computational steps the answer yes to the question if $x \in S$.*

4.4 Church's Thesis

During the early 1930s, the American mathematician Alonzo Church, together with his collaborators Kleene and Rosser, was looking for a *purely logical* way to define computable functions. The calculus he developed is known as λ-calculus.[8] The idea of the λ-calculus is essentially that of a formal system in which rules can be expressed by terms and two basic operations of abstraction and application. The language of this calculus is given as follows:

Definition 48 (The Language of the λ-Calculus) *The syntax of the calculus is given by*

- *Symbols: variables $x_0, x_1, \ldots$; abstraction symbol λ; parentheses and the symbol (.).*
- *Terms:*

  - *a variable is a term*
  - *if M and N are terms, the application (MN) is a term*
  - *if M and x are terms, the abstraction $(\lambda x. M)$ is a term, with x bound in M.*

The calculus has rules to manipulate expressions, given in Figure 4.1. A term is said to be in *normal form* if no further β-reduction can be applied to any of its subterms. A term is *normalizable* if it has a normal form (and not every term has one, as this corresponds to the existence of partial functions). A term is *strongly normalizable* if every of its subterms has a normal form, i.e. if it can be completely reduced to total computable functions. An example of valid expression in this language is:

$$\lambda x. x$$

to indicate the identity function, referring to the argument x of the function returned untouched. Numerals are represented as follows:

---

[8] A history of the λ-calculus, from its origins in combinatory logic to its consolidation and evolution in the typed version, is offered in Hindley and Cardone (2006).

$$\lambda xy . xy = 1$$
$$\lambda xyz . y(xyz) = Succ$$
$$\lambda xy . x \underbrace{(\; \ldots \; (xy) \ldots)}_{n \; times} = n$$

For this calculus, Church and his collaborators were able to prove around 1934 co-extensibility with effective calculability. The first part of this result, given in detail by Kleene, consisted in proving the equivalence of formulas of the λ-calculus with the class of general recursive functions. The first step consisted in proving that all general recursive functions can be defined in the λ-calculus:

Theorem 2 (Kleene (1935, 1936c)) *Every non-negative integral function of natural numbers which is recursive is λ-definable.*

The second step consisted in proving that all functions that can be defined as λ-expressions are recursive:

Theorem 3 (Church (1936); Kleene (1936c)) *Every λ-definable function whose values are well-formed formulae is recursive.*

In the context of this equivalence result, Church intended to translate the informal notion of effective calculability of functions with a formal counterpart. The above result allowed him to state that effective calculable functions, i.e. the class of general recursive functions, are all λ-definable functions. The major result that connects effective computability and recursiveness is then formulated as follows:

Definition 49 (Church's Thesis, Church (1936)) *All effectively computable functions are recursive.*

While the equivalence between recursion and λ-definability was proven, the identity of either of these two notions with the informal notion of computability could not: it was evident that recursion produced computable functions, but it could not be claimed that the informal notion could be exhausted entirely by the available formal counterpart.

The main limitation to this aim was represented by the informal nature of the notion of effective computability: this expressed a natural property of a part of human reasoning, namely that which could be reduced to purely rational processes. But the thesis had much evidence in its favour. In the first place, the existence of two different and equally natural definitions of effective calculability (i.e. recursion and λ-definability) which turned out to be equivalent was a sign that computability denoted a large and possibly unique set of operations that could be expressed in different languages. Moreover, it appeared that no larger class than recursive functions was needed to cover the effectively computable ones: all effectively computable functions and all the methods to produce them are recursive.

At the informal level, recursion coincided with another intution of computability: for any recursive function it exists an *algorithm* to compute any particular value of that function and for λ-definable functions such algorithm consists in the reduction of the formula to its normal form.[9] Soon the intuitive notion of computability as a purely mechanical process would be introduced and it would turn out to satisfy the same equivalence with recursiveness. Algorithmic computation and computability by a machine are the crucial interpretations of recursion at the basis of computing as a discipline in its modern sense.

Exercises

Exercise 21 *List the informal meanings of the axioms necessary and sufficient for defining the standard model of $\mathbb{N}$.*

Exercise 22 *Define addition by recursion from the successor function.*

Exercise 23 *Give an instance of an addition operation constructed by recursion from the successor function.*

Exercise 24 *Define multiplication by recursion from the addition function.*

Exercise 25 *Give an instance of a multiplication operation constructed by recursion from the addition function.*

Exercise 26 *Define exponentiation by recursion from the multiplication function.*

Exercise 27 *Give an instance of an exponentiation operation constructed by recursion from the multiplication function.*

Exercise 28 *Define the basic recursive functions.*

Exercise 29 *Give an instance for each basic recursive function.*

Exercise 30 *Define the composition schema.*

Exercise 31 *Compose the functions $max(x,y)$, $x = 2^z$ and $y = 5 + w$, for some integer value of the variables z, w.*

Exercise 32 *Define the minimization schema.*

Exercise 33 *Consider the function $f(x,y) = x^y$. Is this a primitive recursive function? If so, by which schema?*

Exercise 34 *Define a recursion schema for the inverse of the successor function over $\mathbb{N}$.*

Exercise 35 *Consider the function*

$$x \mathbin{\dot{-}} y = \begin{cases} 0 \text{ if } x = y \\ x - y \text{ otherwise} \end{cases}$$

---

[9] We will reconsider this connection closely in Chapter 6.

Prove that this function is primitive recursive by defining an inductive definition (hint: you can use the inverse of the successor function).

Exercise 36 *Consider the function*

$$x/y = \begin{cases} \text{the least } z \text{ such that } x \leq (y(z+1)) \\ \text{undefined if } x^2 = 0 \end{cases}$$

Which schema does it apply to? Is the function primitive recursive?

Exercise 37 *Consider the exponentiation function x^y. Explain minimization applied to exponentiation: is it regular?, i.e. is there for every input x to exponentiation a value of y such that $x^y = 0$? Is it undefined?*

Exercise 38 *Explain informally the meaning and the limitations of Church's Thesis.*

5 Mechanical Computation

Summary

The previous chapter investigated formally the notion of computable function and the meaning of Church's Thesis. In this chapter we present the model of computation known in the literature as the Turing Machine, we introduce the notion of Universal Machine, and present a translation of computable functions to this model, justifying the Turing Thesis. Finally, we investigate the limits of mechanical computation and its current relevance for the modern understanding of program.

5.1 Turing Computability

Recursive functions and their identity with formulae definable in the λ-calculus illustrate a strict mathematical interpretation of effective calculability. Another way of approaching the representation of effectively computable functions is through the notion of computation by means of a machine. After the engineering work of Charles Babbage (1791–1871) who designed (but never fully constructed) the *Analytical Engine*,[1] a machine meant to use basic arithmetical operations, the mathematical (and so abstract) model of computable processes by a machine was provided by Alan Mathison Turing (1912–1954), in his *On Computable Numbers, with an Application to the Entscheidungsproblem*.[2] We shall first analyse the properties of Turing Machines; then identify the kind of functions and sets (as the domains of those functions) that correspond to the operations on Turing Machines; and finally provide a bridge to import

---

[1] See Babbage (2010).
[2] Turing (1936). One of the most fascinating and complete reconstructions of this convergence of ideas is offered in Gandy (1988).

On the Foundations of Computing. Giuseppe Primiero, Oxford University Press (2020). © Giuseppe Primiero.
DOI: 10.1093/oso/9780198835646.001.0001

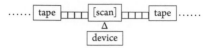

Figure 5.1 Turing Machine

these functions into a logical language with a derivability relation. In this way, on the assumption that all of these forms of computation are equivalent, whatever we shall prove as a property of one, will be proven to hold as a property of the other representation. To this aim, the following is an important remark: by the diagonalization method, we have proven Cantor's Theorem, i.e. that not every set of positive integers can be listed, or in other words that the set P^* of all sets of positive integers is not enumerable; this means that for at least one set (namely P^*) no effectively computable function is definable. If Turing Machines (and in turn other computational representations) will be shown to be a valid encoding for computable functions, it is evident that a listing of such machines becomes possible, and hence an argument holds entirely similar to the one on functions leading to Cantor's Theorem: not every valid Turing Machine can be defined in a list of all Turing Machines.

Turing starts considering human mechanical computability:

> We may compare a man in the process of computing a real number to a machine which is only capable of a finite number of conditions.[3]

An a-machine (for *automatic machine*), or what is known as a Turing Machine in the modern literature, is obtained as a strict regulation of human calculation abilities (a *computor*) to obtain restrictive conditions that would allow for a mechanical definition. This model represents a further—and in some respects more intuitive—form of encoding the notion of computation or calculability, conceptually entirely equivalent to primitive recursive or λ-definable functions.[4] A Turing Machine is thus a *finite automaton*, a mechanical device with *input/output* processes working at discrete times on a finite set of symbols to build a language. The usual intuitive representation of a Turing Machine is of the form given in Figure 5.1. It can be thought of as consisting of a tape (possibly infinite in both direction) scanned by a head, which can read the symbol currently written on the scanned cell of the tape, delete it, write one of the symbols of the given language, move in either direction, halt.

Let us now consider a formal definition.[5]

---

[3] (Turing, 1936, p.231).

[4] See De Mol (2018) for another presentation of Turing Machines and some aspects not considered here like Post's interpretation and the relation with his system to represent computable functions from Post (1936). Computation by machines not as an abstraction to model finite human computation was analysed only in 1980 by Turing's pupil Robin Gandy; see Gandy (1980).

[5] For another presentation of Turing Machines see De Mol (2018).

Definition 50 (Turing Machine) *A Turing Machine (TM) is a model for carrying out mechanical computations on tuples of positive integers. Each computation takes place on a tape divided into squares unending in both directions. Each square may have one symbol {0, 1, ∗} for a binary language and a blank respectively. At each stage of the computation the machine*

- *scans one square of the tape, i.e. it reads its content;*
- *can print a symbol 1 or 0;*
- *can erase a symbol, i.e. it prints a ∗;*
- *can do nothing, i.e. it leaves the current symbol untouched;*
- *moves one square to the left L, or to the right R, or does not move N.*

The machine eventually halts.

Note that in this representation, the TM has a binary alphabet with $1, 0$ as meaningful symbols and ∗ as the empty cell. It is also possible to have a unary language, i.e. where 1 is the only meaningful symbol and ∗ the symbol for the blank cell. The input to the machine is given by a numeral expressed in binary or unary notation on the tape when the machine starts.

At each stage of the computational process, the state of the device is therefore completely and entirely described by one unique *configuration*:

Definition 51 (Configurations of TM) *A TM at each stage of the computation is in one of its internal configurations*

$$Q = \{q_1, \ldots, q_m\}$$

Each $q_i \in Q$ expresses a tuple composed by

$$q_i = (s_i, scan, operation, s_j)$$

defined as follows:

- *the current state s_i*
- *the content of the cell read scan*
- *the operation performed, defined by a*

  - *the symbol written: $1, 0, \ast$*
  - *a movement along the tape: L, R, N*

- *the next state s_j.*

TM is in state s_i when carrying out instruction i. When the instruction is to halt, then no other instruction is given provided the actual state and the next state is H.

The machine has a defined *standard initial configuration*, usually identified with the scanning device being positioned on the first, leftmost entry of the input (of the first input, in case of operations on more than one input). The number of states available to the machine is constrained to a finite number, so that the intuition of an effective computation implemented by a machine of limited complexity is preserved. The finite automaton requires that at each stage of the computation process the description of the state of the machine is *complete*, i.e. that for each state is fully described what happens when in that state the machine is given a certain input from the alphabet. The set of instructions needs also to be *consistent*: given a certain state and symbol read, only one instruction must be possible, or in other words with a state s_i and reading a given symbol, only one operation should be available and only one successive state s_j should be reached. Eventually, the *final configuration* q_m will lead to the final state H and no further state follows. The halting state is the machine's *standard final configuration*, usually identified with the scanning device being positioned on the last, rightmost entry of the output.

A set of configurations for a given Turing Machine is meant to fully and exhaustively expresses the behaviour of that machine, i.e. to represent a *program*:

Definition 52 (Configurations or Program) *The sequence of configurations of a TM says for each stage of the computation what is on the tape at that stage, what state the machine is in at that stage, and which square is being scanned and what the next state is. The full set of configurations for a machine is also called its program.*

The configurations governing the behaviour of the machine was also called by Turing its table of instructions.

Example 8 *A machine M works with an endless blank tape (an empty input) and with the scanner positioned over any square of the tape. M has four states s_1, s_2, s_3, s_4 and is in state s_1 when it starts working. The configurations of the machines (its table of instructions or program) is the following:*

STATE	SCAN	OPERATION	NEXT STATE
s_1	*	$0, R$	s_2
s_2	*	$*, R$	s_3
s_3	*	$1, R$	s_4
s_4	*	$*, R$	s_1

The table is glossed as follows:
State s_1: on scanning a blank square, print 0, move right one square, and go into State s_2.
State s_2: on scanning a blank square, do not print anything, move right one square, and go into State s_3.
State s_3: on scanning a blank square, print 1, move right one square, and go into State s_4.
State s_4: on scanning a blank square, do not print anything, move right one square, and go into State s_1.

Acting in accordance with this program, M when started on a blank tape prints alternating binary digits on the tape, 010101 . . ., working endlessly to the right from its starting place, leaving a blank square in between each digit. Note, however, that no instruction is given either to act if the machine finds a non-blank square, or to halt.

Example 9 *If the initial tape is not empty, instructions needs to be given to the machine for how to act when reading either 1 or 0. If we want to obtain an output that alternates 1, 0 without blanks in between, the following table of instructions suffices:*

STATE	SCAN	OPERATION	NEXT STATE
s_1	*	$1, R$	s_2
s_1	0	$1, R$	s_2
s_1	1	$1, R$	s_2
s_2	*	$0, R$	s_1
s_2	0	$0, R$	s_1
s_2	1	$0, R$	s_1

The program now has a smaller space state of instructions, but a bigger one of possible inputs (as the tape is not empty). This is due to the requirement that a precise representation be given for the calculation to be entirely and completely described at each step. Again, this table is for a program that alternates between two simple instructions and never halts.

It is easy to see that basic computable functions are all Turing computable:

Theorem 4 (Turing Computable Functions) *Basic functions for identity, empty, constant value 1, projection, addition, multiplication, exponentiation, super-exponentiation, and so on are all Turing computable.*

This result is easily proven by induction, showing that one such machine exists for each such function.

Example 10 *Consider a TM using binary arithmetics for the identity function. Its construction is left as an exercise to the reader at the end of this chapter.*

Example 11 *Consider TMs using unary arithmetic for the empty and constant value 1 functions. Their construction is left as an exercise for the reader at the end of this chapter.*

Example 12 *Consider the representation by a TM using unary arithmetic of the arithmetical operation of sum $p + q$ of two blocks of entries of 1s. In the case of $p = 2$ and $q = 3$, it means to go from a tape of the form*

$$\cdots *11 * 111 * * \cdots$$

to a tape of the form

$$\cdots * 11111 * * \cdots$$

The computation can be obtained in two different ways:

1. *going to the leftmost 1 of the left block p, deleting it and rewriting it on the right-hand side of the last 1 of the second block q, and so for as many times as many 1s there are in block p (i.e. in this case twice);*

2. *removing the leftmost 1 of the left block p, deleting it and rewriting it on the * separating the block p from the second block q of 1s.*

The construction of the appropriate instruction tables is left at the end of this chapter as an exercise for the reader.

Example 13 *Consider the arithmetical operation n · 2, with a unary encoding for each n: informally, one way to describe the machine performing this operation is by saying that the scanning device will read each entry of the input, delete it, and write it twice after the end of the whole input. The application of a well-defined set of configurations for this program with n = 2 will allow the machine to transform a tape of the form:*

$$\cdots * *11 * *\ldots$$

to a tape of the form

$$\cdots * *1111 * *\ldots$$

Such a program can be informally defined by the following operations:

1. *start from the leftmost non-empty cell, read its contents, and delete it;*

2. *go to the first rightmost empty cell and move right;*

3. *write a 1, move right, write a 1;*

4. *move left across the output, the empty cell, the input until the first entry of the input is found;*

5. *repeat from instruction 1;*

6. *perform this set of operations until no input is left, then halt.*

The following table of instructions suffices to implement this function:

STATE	SCAN	OPERATION	NEXT STATE
s_1	1	$*, R$	s_2
s_1	$*$	$*, N$	H
s_2	1	$1, R$	s_2
s_2	$*$	$*, R$	s_3
s_3	1	$1, R$	s_3
s_3	$*$	$1, R$	s_4

STATE	SCAN	OPERATION	NEXT STATE
s_4	1	$1, R$	s_4
s_4	$*$	$1, L$	s_5
s_5	1	$1, L$	s_5
s_5	$*$	$*, L$	s_6
s_6	1	$1, L$	s_6
s_6	$*$	$*, R$	s_1

Note that in this table all configurations are defined only for inputs $1, *$.

Let us follow the execution of this program when the machine is given input $n = 2$; the intended output is 4. The machine starts in state s_1 reading the first 1, it overwrites it with a $*$, and moves to the right entering state s_2; in this state the machine will read a 1, do nothing on it, and move right in the same state; then it will read the symbol $*$ at the end of the input, it will leave it and move to the right, entering state s_3; in s_3, the machine will scan a symbol $*$, overwrite it with 1, move to the right, and entering state s_4, where the same overwriting of $*$ with 1 will happen and then the machine will start moving left, entering state s_5; in state s_5 the machine will go through the tape, leaving everything untouched until it reaches the symbol $*$ at the beginning of the output, when it will keep moving left but in state s_6; in this new state, the head moves leftwards keeping everything untouched until it reaches the beginning of the input, when it will start moving back to the right in state s_1. From this point on the entire operation will be repeated for the second input entry 1. When also this one will be deleted and two entries placed at the rightmost side of the tape, the tape will re-enter state s_6 and successively state s_1, when it will encounter a blank $*$ and finally halt.

Example 14 Consider the representation by a TM of the arithmetical operation of multiplication of $p \cdot q$ of two blocks of respectively p and q entries 1: for $p = 2$ and $q = 3$, it means to go from a tape of the form

$$\cdots * 11 * 111 * * \ldots$$

to a tape of the form

$$\cdots * * 111111 * * \ldots$$

Let the machine start from the leftmost stroke of the left block p, let it count how many strokes are in p, if $p = 1$ then $p \cdot q = q$ and the machine deletes the stroke of the p block and halts at the rightmost stroke of the q block; if $p = n > 1$, then the machine counts such n, goes to the rightmost stroke of the q block, and moves it of q squares; then it moves back to the one stroke before the rightmost and move it of q squares, and so on for all strokes in the q block; the machine shall repeat this operation $n - 1$ times, each time deleting the leftmost stroke from the p block; when the p block amounts to 1, the machine will fill with strokes all the squares left between the leftmost of the q block and the remaining stroke of the q block and *then it halts at the rightmost stroke of the q block. The construction of the relevant instruction table is left at the end of this chapter as an exercise for the reader.*

These are examples of numerical functions that are *computable by a Turing Machine*. To provide a general definition, we need to abstract with respect to the number of arguments the function can be applied to:

Definition 53 (Numerical function computable by a TM) *A numerical function f of k arguments $m_1, \ldots, m_k$ is Turing computable if there is some TM that computes it, in the following sense:*

1. *the k arguments of the function are represented by k blocks, each encoding in an appropriate language (e.g. binary or unary) the value of each m_i and separated one from another by a blank; TM starts in the standard initial configuration;*

2. *TM halts on a tape with a block encoding the value n equal to the result of the function applied to the k arguments; TM ends in the standard final configuration when there is no further instruction to be performed;*

3. *if the program expressing f has no value for the arguments $m_1, \ldots, m_k$ represented initially on the tape, TM will never halt, or will halt in non-standard configuration.*

In the appendix to his seminal work, Turing shows that computability by TMs and computability by λ-definability are equivalent. The possibility of establishing a generalized identity for the various forms of computability relies on their identity with the notion of *general recursiveness*. We have mentioned that the basic computable functions are Turing computable (Theorem 4); in the following we will also show that the computational schemas preserving computability can be still formulated in terms of TMs. To this aim, it is enough to show that the Composition schema from Definition 39, the Primitive Recursion schema from Definition 40 and the Minimization schema from Definition 42 can all be interpreted as TMs.[6]

Theorem 5 (Composition as TM) *Given a function*

$$h(x_1, \ldots, x_n) = f(g_1(x_1, \ldots, x_n), \ldots, g_m(x_1, \ldots, x_n))$$

with f and g computed by machines $M_1, \ldots, M_m, M_{m+1}$. Then there is a machine M implementing h.

Proof. The proof can be sketched as follows. Write on the tape copies for the inputs for $M_1, \ldots, M_m, M_{m+1}$; simulate M_1, remove the copy of its input, and write its output to the rightmost empty cell of the inputs. Repeat the above for the machine M_2 and its input, and do so for all the machines up to M_m. Now simulate M_{m+1}, remove all the outputs from the previous simulated machines, and move the output of M_{m+1} near the inputs. □

---

[6] These results are adapted from Oddifreddi (1992).

Theorem 6 (Primitive Recursion as TM) *Given a function*

$$h(x,0) = f(x)$$
$$h(x,y\prime) = g(x,y,h(x,y))$$

with f and g computed by machines M_1, M_2. Then there is a machine M implementing h.

Proof. The proof can be sketched as follows. Write on the tape after the inputs copies s for y (a counter to be decreased at each step), copies of the inputs and an empty string t; simulate M_1, if our counter is not exhausted, increase t by a 1, decrease s by a 1, simulate M_2, and move it to the previous place of the output of M_1. Repeat the above until s is empty, then remove all the outputs from the previous simulated machines, and move the output near the inputs. □

Theorem 7 (Minimization as TM) *Given a function*

$$\mu[f(x_1,\ldots,x_n)] = \begin{cases} y, & \text{if } f(x_1,\ldots,x_n,y) = 0 \text{ and for all } t < y \\ & f(x_1,\ldots,x_n,t) \text{ is defined and } \neq 0 \\ \text{undefined} & \text{if there is no such } y \end{cases}$$

with f computed by a machine M. Then there is a machine $M\prime$ implementing μ.

Proof. The proof can be sketched as follows. Write on the tape copies of the inputs and a string y with one 1; simulate M, if the output is not 0, erase the output, increase y by a 1, and simulate M again. When the output is 0, then remove all the inputs, the output, and write the value of y near the inputs. □

As a corollary from the above, we obtain the following:

Theorem 8 (Turing (1936)) *Any recursive function is Turing computable.*

5.2 The Universal Machine

The TMs introduced above are hard-wired computers, made to compute one kind of function on any given input. The idea of a TM that could run *any* program was also introduced by Turing. The *universal computing machine,* now known as the Universal Turing Machine, can be programmed to carry out any calculation. In Turing's words:

> It is possible to invent a single machine which can be used to compute any computable sequence.[7]

The Universal Machine U has a single, fixed table of instructions built into it. Operating in accordance with this one fixed table, U can read and execute coded instructions inscribed on its tape. These correspond to the program of any specific machine M that U wants to emulate:

[7] (Turing, 1936, sec.6, p.241).

If this machine *U* is supplied with a tape on the beginning of which is written the standard description of some computing machine *M*, then *U* will compute the same sequence as *M*.[8]

With this intuition Turing provides the translation of program as code and gives one of the earliest models of the 'stored program' concept, the idea of controlling the function of the computing machine by storing a program of instructions in the machine's memory.[9] The Universal Machine *U* requires an actual input: on its tape there will be the program of a machine *M* describing all of its configurations and, following these instructions, there will be the input *i* to be fed to *M* executed by *U*. The machine *U* reads and interprets the program of *M* on input *i* and prints the same output *o* that the machine *M* would print on input *i*. The output *o* must be written on the tape after the program of *M* and the input *i*, to avoid overwriting it. Different programs of different machines can be inscribed on the tape, enabling the UTM to carry out any task for which a Turing Machine instruction table can be written. Thus, a single machine of fixed structure is able to carry out every computation that can be carried out by any Turing Machine whatsoever.

The Universal Turing Machine requires therefore an encoding of both program and input on its tape. Take languages over alphabets containing at least 2 symbols $L = \{0, 1\}$. It is always possible to encode a string from a large alphabet using a string with just the alphabet of L. For example:

1. devise an alphabet with a character for each element of a program;
2. build a fixed-length binary encoding for each symbol in some order (e.g. keywords first, blanks second, rest of input after that);
3. substitute each character in the program with its binary encoding.

Given a TM program denoted by *P*, now $\langle P \rangle$ denotes its encoding. For example, the following encoding can be used for a program that requires a single state:

$$s = 1$$
$$1 = 10$$
$$0 = 01$$
$$R = 001$$
$$L = 100$$
$$N = 000$$
$$* = 111$$
$$e = 0$$

---

[8] (Turing, 1936, sec.6, pp.241–2).

[9] For the debate around the origins of the concept of stored program and its different interpretations, see Chapter 8.4.

Then, the following program to add a 1 at the end of a string

STATE	SCAN	OPERATION	NEXT STATE
s_1	1	$1, R$	s_1
s_1	0	$1, R$	s_1
s_1	*	$*, N$	H

could be encoded as the three strings

$$110100011$$
$$101100011$$
$$11111110000$$

We could also decide to have one blank to encode each beginning of a new program line. Now we can encode also the input, maybe just by the identity function, so that

$$1 = 1$$
$$0 = 0$$

We could also decide to have two blanks to encode the separation of the program from its input. Now a program P with input i can all be encoded as $\langle P, i \rangle$:

$$110100011 * 101100011 * 11111110000 * *1$$

This will print somewhere else on the tape (for example after three blanks) the result:

$$110100011 * 101100011 * 11111110000 * *10 * * * 11$$

This construction is entirely general and can be obtained for any set of configurations and any input, hence it can be generalized:

Theorem 9 (Turing (1936)) *There is a Turing Machine UTM called the Universal Turing Machine that when run on $\langle P, i \rangle$ simulates the Turing Machine for program P on input i.*

The Universal Turing Machine is an essential step, from a conceptual, formal, and philosophical viewpoint. In the first place, the UTM denotes the evolution of a model of computation from a design that allows one to implement any specific computable function in a mechanical process, to one which can implement *any* such function. Secondly, it has been argued elsewhere that the UTM actually expresses the theoretical origin for the 'stored-program' model of computers: it appears obvious that the idea of the memory (the tape) of a Universal Machine U used to store the program of any special purpose machine M has a close conceptual (if not necessarily technical) resemblance to the engineering principle of the stored program; nonetheless the latter would be realized in a different context, independently of Turing's work. Finally, and maybe more significantly for the mathematical foundation of computing, the UTM is in itself a justification of the Church-Turing Thesis, as it provides a model of computation sufficiently powerful to simulate any other computation, without requiring stepping outside the limit of one model of effective computability already proven equivalent to general recursiveness.

5.3 The Halting Problem

The problem of determining whether all functions are computable has been anticipated by Theorem 1 in Chapter 3 as the existence of a nonenumerable set, i.e. that the power set of a countably infinite set is uncountably infinite. By Theorem 8, this corresponds to nonenumerability of the set of all Turing computable functions: the latter statement can be intuitively understood by considering that each TM can be represented by a string of symbols, and therefore one can list them forming an enumerable set of such machines. Any possible representation of TMs which defines the tuples of the configuration in terms of finite strings of symbols can be encoded, and therefore their entire list can be enumerated. Hence, one could ask for a TM that lists all such encodings: is this machine then in the set of all TMs or not? According to an application of Cantor's Theorem, there must exist functions that are not Turing computable.

To show that there are non-Turing computable functions, one defines a function such that supposing it to be in any place of the list of TMs would produce a contradiction. This is obtained by an application of Cantor's diagonal method.[10]

Definition 54 (The Diagonal Function) *The diagonal function d is defined as:*

$$d(n) = \begin{cases} 0, \text{if } f_m(n) \text{ is defined and } = 1 \\ 1, \text{ otherwise} \end{cases}$$

This function is a total function of one argument, such that

- when applied to argument n it gives value 0 when the function $f_m(n)$ defining the mth TM running on argument n is defined and it gives value 1;
- and it gives value 1 otherwise.

Then the following can be shown:

Theorem 10 *The diagonal function d is not Turing computable.*

Proof. Suppose d is the computable function implemented by the mth Turing Machine in the list of all TMs, then for each positive integer n either $d(n)$ and $f_m(n)$ are defined and equal, or none of them is. Consider now the case $m = n$, in this case whether $f_n(n)$ is defined or not, one obtains a contradiction:

---

[10] For this proof, see Boolos et al. (2002).

$$f_n(n) = d(n) = \begin{cases} 0, & \text{if } f_n(n) \text{ is defined and } = 1 \\ 1, & \text{otherwise} \end{cases}$$

1. If $f_n(n)$ is undefined or it is defined but it has value $\neq 1$, then $d(n)$ will have value 1 and because of the identity $f_n(n) = d(n)$, then also $f_n(n)$ will have value 1 (but this contradicts the previous claim that $f_n(n) \neq 1$); but so, if $f_n(n) = 1$, then it is defined (which contradicts the initial assumption that $f_n(n)$ is undefined);

2. If $f_n(n)$ is defined and has value 1, then $d(n) = 0$ and because $f_n(n) = d(n)$, then $f_n(n) = 0$, which contradicts the assumption.

In both cases, from the assumption that d appears at some place n in the list of Turing computable functions we have obtained a contradiction, thus the supposition must be false. $\qquad\qquad\square$

According to the definition of $d(n)$, one can establish how the machine M_n computes by following its operations from the starting configuration up to its final configuration: when it halts one can establish if it is defined, i.e. if it halts with value 0. But, obviously, one way $d(n)$ can get value 1 is if M_n never halts at all, because then it means that the corresponding function $d(n)$ is not defined, which happens if the machine obtained by such function produces no value when started with input n.

Hence, the general problem of non-Turing computable functions can be reformulated as the halting problem for TMs:

Definition 55 (Halting Problem) *Is it possible to determine in a finite amount of time whether machine M_n started scanning in standard configuration does or does not eventually halt? The problem consists in finding an effective procedure that given any machine M represented by its function f and given any number n will say whether or not machine M with input n ever halts.*

The halting problem can be formulated in terms of a corresponding function.

Definition 56 (Halting Function) *The halting function $h(m, n)$ of two arguments is the function of machine m starting with input n. The problem expressed by this function is if machine m with input n halts (provided m, n can take any values, this turns out to be precisely our halting problem). The halting function is defined as follows:*

$$h(m,n) = \begin{cases} 1, & \text{if machine } m \text{ with input } n \text{ halts;} \\ 0, & \text{otherwise.} \end{cases}$$

If h is computable, then d is effectively computable. But d is not computable, and then h cannot be computable. As this function turns out to be non-computable, then the problem cannot be solved.

Theorem 11 *The halting function h is not Turing computable.*

Proof. Take a number for a machine n; then one can compute $h(n, n)$, that is machine n with input n.

1. If $h(n, n) = 0$, it means machine n on input n does not halt: then the corresponding function $f_n(n)$ is not defined and therefore the corresponding diagonal function $d(n) = 1$. This in turn means that on the assumption that $f_n(n) = d(n)$, then also $f_n(n) = 1$ and it is defined, so $h(n, n)$ is also defined and the machine n on input n has to stop (contradiction);

2. If machine n on input n halts then $h(n, n) = 1$:

   - if it halts in non-standard configuration, then $d(n) = 1$ (by definition of $f_n(n)$); but because $f_n(n) = 1$ then $d(n) = 0$, by Definition 54;

   - if it halts in standard configuration and $f_n(n) \neq 1$ then still $d(n) = 1$; a contradiction arises as above;

   - if it halts in standard configuration and $f_n(n) = 1$ then $d(n) = 0$ again by Definition 54, and again a contradiction arises.

But because d is not Turing computable, assuming Turing's thesis it follows that d is not effectively computable and therefore h is not effectively computable and therefore not Turing computable.

$\square$

5.4 Turing's Thesis

The reduction of general recursiveness to Turing computability is obviously a further building block to prove the generality of recursion as a precise expression of effective computability. Recall that by Church's Thesis (Definition 49), all effectively computable functions are recursive. Provided recursiveness implies Turing computability the nature of the Turing Machine can be generalized, implying functional preservation even under relaxed hardware conditions:

Theorem 12 (Shannon (1956)) *The same class of functions are Turing computable whether one defines Turing Machines to have a tape infinite on both directions or infinite in only one direction; whether one requires Turing Machines to operate with one symbol in addition to the blank or to operate with any finite number of symbols; and whether the machine has only a fixed number of states $n \geq 2$, or any number of states.*

The different forms of computations are supposed to include all and the same set of functions that are computable in the intuitive sense of effective computability. By Church's Thesis, we have established the identity between recursiveness and computability, thus categorizing intuitively effectively computable functions in terms of recursive functions. This has been generalized to sets and relations, relying on the properties of the related characteristic functions and their closure by logical operations. On the other hand, by Theorem 8, every recursive function is computable by a Turing Machine. In order to complete the identity that relates our different notions of computability, we now need to prove the identity between Turing computability and recursiveness:

Theorem 13 *Every function computable by a Turing Machine is recursive.*

Proof. The proof consists in interpreting every step that can be performed by a TM by means of a recursive function through a coding operation on the content of the tape.[11] The coding includes functions for the machine halting in standard *stdh* and non-standard *nstdh* position with a given output *o* at time *t*. The function expressing if and when the machine *m* is going to halt on input *x* uses minimization, as it determines the minimal time-value *t* such that the machine halts in its standard configuration:

$$halt(m, x) = \begin{cases} \text{the least } t \text{ s.t. } stdh(m, x, t) = 0 \text{ if such a } t \text{ exists} \\ \text{undefined otherwise} \end{cases}$$

By using minimization, the representation of TMs cannot be constrained to primitive recursive functions. In turn, the halting function identifies a semi-decidable relation: there is a value for it if the machine halts; if the machine does not halt, no such value can be given. □

We can now establish the desired result:

Theorem 14 *A function is recursive if and only if it is Turing computable.*

Proof. Let $F(m, x) = otpt(m, x, halt(m, x))$ be a recursive function that gives the output value of TM with code number *m* for argument *x*. Such a function *F* is construed by recursive functions and it remains undefined if no output is given by *m* on input *x*. For any code *m* of a TM computing a function f, $f(x) = F(m, x)$; since *F* is recursive, *f* is recursive as well. □

Using the identity expressed by Church's Thesis, we are now in the position to formulate the following:

Definition 57 (Turing's Thesis) *All effectively computable functions are Turing computable.*

This version of the thesis suffers from the same limitations of its counterpart as formulated by Church: the notion of effective computability is in itself an informal notion and only the formal notion of recursiveness can be shown to be identical to the formally defined notion of a TM.

Recursion provides therefore a mathematical explanation of what computability means. In turn, it also offers a characterization of the effective method required by the decision problem (cf. Proposition 2, Ch.3.4):

---

[11] Such a translation is provided in detail in (Boolos et al., 2002, ch.8) via a Wang code.

Definition 58 (Decision Problem (Computable Version)) *Is there a recursive function* $f(\Gamma, \phi)$ *such that given a finite set of sentences* Γ *and a sentence* ϕ *within a system of logic* F *would establish whether* $\Gamma \vDash_F \phi$?

This means to reduce the problem of deciding whether a formula is a consequence of a set of premises to an arithmetical problem of the following form:

$$f(\Gamma, \phi) = \begin{cases} 1 \text{ iff } \Gamma \vDash_F \phi \\ \text{undefined otherwise} \end{cases}$$

Note that the second case reflects the fact that the consequence set of Γ is only semi-decidable in general.

We have also suggested (see Proposition 3) that the effective method of the decision problem has an equivalent formulation in terms of a mechanical procedure, where Γ is a machine (now interpreted precisely as a TM equipped with an input ι on its tape), ϕ is an output, and the system of reference F is the set of configurations of the machine Γ expressing its program, which can be encoded and then translated in first order logic form. This language is logically and functionally equivalent to a program written in a general purpose language L. Then the result of Theorem 14 gives us the following:[12]

Theorem 15 (Recursive Functions as Programs) *Every primitive recursive function can be mechanically computed by a program P written in any general purpose programming language L.*

Proof. The proof goes by induction:

Base case: the only primitive recursive functions that can be derived with only one step are the functions successor, projection and zero. For any of these we can define—say—a program in L, by Theorem 4.

Induction Step: Let $f(x_1, \ldots, x_n)$ be a function that is primitive recursive by a derivation of $m + 1$ steps. If the last rule used in the derivation of this function is either the zero rule, projection rule, or successor rule then $f(x_1, \ldots, x_n)$ is one of those functions, and thus is primitive recursive. If the last rule used was composition then there are functions

$$g_1(x_1, \ldots, x_n), \ldots, g_k(x_1, \ldots, x_n), h(x_1, \ldots, x_k)$$

which are primitive recursive such that each of those functions takes $\leq m$ steps to derive, and

$$f(x_1, \ldots, x_n) = h(g_1(x_1, \ldots, x_n), \ldots, gk(x_1, \ldots, x_n)))$$

Since each of the g_i and h take m or less steps to derive, by the induction hypothesis they can be computed by a L program (i.e. by applying Theorem 5). If

---

[12] A similar argument is developed for a specific language in Reus (2016).

the last rule used was primitive recursion then there is a function as by Theorem 6 for its computation as a program P in L.

□

The class of effectively computable functions can be further extended, going beyond the set of those computable functions that are total, i.e. defined over any possible input, and including partial functions, i.e. computable only on *some* inputs. Let us give first an intuitive definition of effectively computable partial functions for programs:

Definition 59 (Partial Functions computable by a Program) *A partial function is effectively computable if a list of definite explicit instructions can be given as a program P in L by which,*

- *P arrives in a finite number of steps at the value of $f(x)$ if the function is applied to an argument x in its domain;*
- *P goes on forever without obtaining a result if the function is applied to any argument x outside of its domain.*

Accordingly, the halting problem can be redefined as a program P in any language L. Let us start with an adapted definition of the problem:

Definition 60 (Halting Problem for Programs) *Given a program P in some language L and value i, does program P terminate when given i as input?*

A simple example for such a L language is just one for addition of unary numbers as by Figure 5.2.

This program P reads the input list i by looking through its entries one by one: while reading the first entry, it adds it to the second, then recursively calling the first step. When the input is empty the first step is no longer true, then it halts. We can now define an appropriate uncomputable function:

```
addition read input {
X := first L;          // X is the first argument m in
                          the input
Y := first (rest L);   // Y is the first of the rest of
                          the input
                       // i.e.\ the second argument n

while X {                 // run through X
        Y := compose empty input with Y; // Y := Y+1
        X := rest X       // X := X-1
}
}
halt
```
Figure 5.2 Program for the addition of unary numbers

```
diag read X {
A  := [ X,X ];                 // diagonalisation
B  := halt[ X,X ];             //
Y  := store B;

while Y {                      // run through Y
Y  := Y;          // Y := Y+1
}
}
output Y                       // Y
```

Figure 5.3 Program for the halting function

$$halt(a) = \begin{cases} true, \text{ if } a = [p,i] \text{ and } p(i) terminates \\ false, \text{ otherwise} \end{cases}$$

A program *diag* to show that *halt(a)* is not computable is shown in Figure 5.3. Does the program in L corresponding to *halt* terminate when *halt(halt)* runs? The strategy to show that this question cannot be answered is as follows:

- call $A := halt(halt)$
- execute $B := halt[halt(halt)]$
- execute $Y := B$ and compose it to Y

  1. while $Y := true$, then one enters in *halt* a non-terminating loop where Y calls itself; but if *halt(halt)* $:= true$ by virtue of the former, then it terminates (by definition) and hence there cannot be a non-terminating program, so we get a contradiction;

  2. if $Y := false$, then *halt(halt)* does not go into the loop and Y must therefore terminate; hence it cannot call itself but by assumption if $Y := false$ it means *halt(halt)* $:= false$ and this means *halt* does not terminate, hence another contradiction.

According to the above result, we can now redefine decidability and semi-decidability for Turing Machines:

Definition 61 (Decidable Language) *A programming language L is said to be decidable if there is a Turing Machine that computes a valid output for every string in L accepted as input and rejects every string not in L.*

Definition 62 (Semi-Decidable Language) *A programming language L is said to be semi-decidable (or computably enumerable, or recursively enumerable) if there is a Turing*

Machine that computes a valid output for every string in L accepted as input and rejects or loops on every string not in L.

Theorem 16 (Semi-Decidable Halting Problem) *The halting problem is semi-decidable in any programming language.*

Proof. The proof follows from the definition of semi-decidability and the construction of the program *halt*. ☐

The identity of effective computability with λ-definability and Turing computability, and in turn the corresponding equivalence of the latter two formal notions, has become known in the literature as the *Church-Turing Thesis*. We have already stressed that the thesis, as it refers to the informal notion of *effective computability*, is in principle accepted but considered unprovable.[13] The reasons that have supported its validity can be summarized as follows:

1. for every effectively calculable function known it is possible to provide the table of configurations of a Turing Machine to compute it;

2. all construction schemas for computable functions from computable functions have counterpart methods by constructions of appropriate Turing Machines;

3. every formal translation of the notion of effective computability has identified the same class of functions, namely those that can be computed by a Turing Machine.

There have been nonetheless some attempts at formalizing the thesis. These approaches try to render formally the notion of effective computability in some independent way from the other terms of the equivalence (i.e. respectively λ-definability and computability by a Turing Machine).[14] A first way to do this[15] is to eliminate internal states of Turing Machines in favour of physical counterparts and to replace the representation of Turing Machines by Post's production systems. This provides a model of effective computability with the following constraints:

Principle 1 (Boundedness, Sieg (2008)) *A computer can immediately recognize only a bounded number of configurations.*

Principle 2 (Locality, Sieg (2008)) *A computer can change only immediately recognizable configurations.*

As a consequence, the system is deterministic, in that any given recognized subconfiguration determines uniquely the next computation step. These conditions are not valid in all systems that interpret effective computability according to Church's Thesis: for example, this is the case for Gödel equational calculus, where substitution operations involve terms of arbitrary complexity; and in the model based on Turing Machines equivalent to one where modifications can happen in parallel on arbitrarily many bounded

---

[13] This position is traced back usually to (Kleene, 1952, sec.62, pp.317–23).

[14] This approach reflects 'argument I' in Turing (1936).

[15] Proposed in Sieg (2002, 2008).

parts (through translation to the model of Gandy machines, which notoriously allow for parallel computations[16]). But a Turing computor (intended as a model of effective calculability) has to respect the locality and boundedness constraints.

A second approach consists in axiomatizing effective computability by the following postulates:[17]

1. An algorithm determines a sequence of computational states for each valid input.

2. The states of a computational sequence are structures. And everything is invariant under isomorphism.

3. The transitions from state to state in computational sequences are governable by some fixed, finite description.

4. Only undeniably computable operations are available in initial states.

The class of processes characterized by these postulates is one of deterministic and sequential algorithms, not restricted to computable functions and not admitting uncomputable oracles. The computational model of abstract state machines satisfies these postulates.[18] Moreover, it characterizes effective computability in terms of abstractness, boundedness, arithmetical effectivity and it includes all partial recursive functions. In view of this interpretation, Church's Thesis states that every numeric (partial) function computed by an arithmetical algorithm is (partial) recursive (the inverse is true by the stated satisfaction of the postulates by abstract state machines). A formally provable variation of Church's Thesis states that every numeric (partial) function computed by an arithmetized algorithm is (partial) recursive. To put it informally:

> No matter what additional data structures an algorithm has at its disposal, it cannot compute any non-recursive numeric functions, since essentially the same computations can be performed over the natural number.[19]

A third strategy has been to identify computation with mathematical deduction from a finite set of instructions expressible in a first-order language.[20] The argument relies on the following well-known result by Kurt Gödel:[21]

Theorem 17 (Completeness) *If a formula ϕ is logically valid in a formal language F, denoted $\models_F \phi$, then there is a finite deduction (a formal proof) of the formula ϕ in F, denoted $\vdash_F \phi$.*

---

[16] See Gandy (1980).

[17] For this approach, see Dershowitz and Gurevich (2008).

[18] See Gurevich (1993, 2000).

[19] Cf. (Dershowitz and Gurevich, 2008, p.338).

[20] This approach, which reflects 'argument II' in Turing (1936), and which distinguishes finitely derivability by rules from strictly mechanical computation, is presented in Kripke (2013).

[21] See Gödel (1929).

This is generalized to derivability and consequence from a set of sentences, respectively $\Gamma \vdash_F \phi$ and $\Gamma \vDash_F \phi$. We know from previous results that given the recursiveness of the derivability relation $\vdash$, when ϕ is of the form $f(a)$, i.e. a function, then it must be a recursive one (and hence Turing computable). It follows that any function whose graph is computable by a deductive argument formulated in a first-order language must also be recursive. This means that for algorithms whose instructions and steps can be formulated in a first-order language, the Church-Turing characterization of the class of computable functions is a corollary of the Gödel Completeness Theorem.

In these attempts to characterize the notion of effective computability in order to prove the Church-Turing Thesis, a common and basic notion seems to occur as the definitional basis of computation, to play the role of a normalized translation of effective computability: the notion of algorithm. What an algorithm is, and the technical and philosophical debate around its nature, are the objects of investigation of Chapter 6.

Exercises

Exercise 39 *Define informally the essential characteristics and structure of a Turing Machine.*

Exercise 40 *Define formally what is the configuration of a Turing Machine and what is a program for such a machine.*

Exercise 41 *Which primitive recursive functions are Turing computable? Which function(s) of Turing Machine require(s) extension to partial recursive functions and why?*

Exercise 42 *Provide the configuration table of a TM that alternates $1, 0$ with interleaving blanks starting from any input tape (i.e. not necessarily blank).*

Exercise 43 *Write the program for a Turing Machine to implement the function in Example 10.*

Exercise 44 *Write the program for a Turing Machine to implement the function in Example 11.*

Exercise 45 *Write the program for a Turing Machine to implement the function in Example 12.*

Exercise 46 *Write the program for a Turing Machine to implement the function in Example 14.*

Exercise 47 *What is the principle of program-as-code and how is it implemented in the Universal Machine?*

Exercise 48 *Explain why the diagonal function cannot be computed by any Turing Machine. Which function intuitively computable in mechanical terms corresponds to the diagonal function and what are the consequences for mechanical computability at large?*

Exercise 49 *Explain the relations between recursiveness, effective computability, and Turing computability.*

Exercise 50 *Explain the relation between the decision problem in logic and the halting problem in the theory of computability by Turing. Explain how the former is affected by the uncomputability of the halting function.*

Exercise 51 *Is it possible to write a program that given any routine and every input tells us whether that routine accepts that input and provides a valid output?*

Exercise 52 *Can you tell if the following program halts or not? Why?*

```
1   def f(int n)
2       if (n == 0)
3           return 1
4       else return n*(f(n-1))
5       end
6   end
```

Exercise 53 *Can you tell if the following program halts or not? Why?*

```
1   def f(int n)
2       if (n < 0)
3           f(n + 2)
4       elsif (n > 0)
5           f(n - 2)
6       end
7   end
```

Exercise 54 *Can you tell if the following program halts or not on any input? Try with some large input, e.g. 156:*

```
1   void f(int x) {
2       while (x > 1) {
3           if (x % 2 == 1) x = 3*x + 1;
4           else            x = x / 2;
5       }
6   }
```

Exercise 55 *Consider the attempts at proving the Church-Turing Thesis: which strategies do they use and how do they differ?*

6 On the Nature of Algorithms

Summary

In the previous chapter, we presented the general notion of computation in its mechanical interpretation and introduced the associated notion of program. In the present chapter, we generalize this analysis in the larger context of the notion of *algorithm*. While general recursion and its different interpretations are a product of mathematical research of the nineteenth and early twentieth centuries, much older is the idea that knowledge should be based on finite and definite procedures. Despite these ancient origins, the nature of algorithms and how their formal and conceptual definition should be given are still debated today.

6.1 Fast Backwards

The history of algorithms spans all cultures and traditions, and their use and study is in the domain of several disciplines, from mathematics to geometry to computer science. It is not the aim of this chapter to offer an overview of this history,[1] but rather to reflect on the conceptual nature of algorithms as they emerged from the foundational debate in mathematics and from the mechanical understanding of computing illustrated in the previous chapters. Nonetheless, a brief historical overview is useful to establish some aspects of interest to us.

The word algorithm comes notoriously from the name of the ninth-century Persian mathematician Muhammad ibn Musa al-Khwarizmi. He gave explicit solutions to linear and quadratic equations of the following form:

---

[1] This task largely exhausted in Chabert et al. (1999).

On the Foundations of Computing. Giuseppe Primiero, Oxford University Press (2020). © Giuseppe Primiero.
DOI: 10.1093/oso/9780198835646.001.0001

$$ax^2 + by + c = 0 \text{ is solved by } \frac{-b \pm \sqrt{b^2 - 4ac}}{2a}$$

and was also responsible for the diffusion of the Hindu-Arabic numerical system in the Middle East and in Europe through his *On the Calculation with Hindu Numerals*, translated into Latin as *Algoritmi de numero Indorum*.[2]

Before him, the Ancient Greeks were obsessed with various mathematical problems, like those of squaring the circle or bisecting an angle, that would be solved through instruments and sets of instructions. Let us consider angle bisection as an example, requiring only the use of straight-edge and compass:

1. Take an angle given by two intersecting straight lines.
2. Using a compass, draw a circle of arbitrary radius from the point of intersection.
3. Using a compass, draw circles of the same radius from the two points where the previous circle hits the straight lines.
4. Draw a line through the intersections of these circles (and the intersection of the straight line).

The original angle is now bisected, although these instructions cannot be used to define angle trisection.[3] In a similar vein, the Greeks gave many sets of instructions for other geometric constructions.

A first characteristic of algorithms intended in either of the two senses above concerns their presentation. The algorithmic style of mathematics was developed by the Ancient Greeks along with its demonstrative counterpart: the former followed the tradition more known to Babylonians and Egyptians, the latter was represented by the style established and known from Euclid onward. In the algorithmic style of mathematical proof, the subproblems composing the task for which an algorithm is formulated are sequentially computed with the intrinsic possibility of being re-used in other combinations, a feature that will return in its modern interpretation as sub-routines of a program. In their ancient formats, algorithms were textually or graphically presented. The repeatability and compositionality aspects of algorithms are crucial both in terms of clarity and comprehension. The algorithmic style of mathematical knowledge construction has resulted as more appropriate in the context of practical problems solved by the use of a tool or instrument.[4] For example, the Antikythera calculated astronomical phenomena against a calendar that included, e.g., the Olympic Games: it is usually considered an analogue computer but its state space is discrete. The best-known analogue computer, with continuous state space, is the slide rule which allows one to multiply and divide numbers by using logarithms:

---

[2] For an English translation, see Crossley and Henry (1990).
[3] Angle trisection was proved in general impossible, using Galois Theory, by Pierre Wantzel in 1837.
[4] This aspect is strongly stressed in Bullynck (2015).

$$x \times y = z \Leftrightarrow \log x + \log y = \log z$$

Other problems treated by analogue mechanisms in the following centuries were tides prediction, ballistics, and flight control on aircrafts.[5]

We identify thus two essential characteristics of algorithms: *language* and *implementation*. Both aspects were strongly stressed by the mechanical interpretation of computing introduced by the Turing Machine model: first, the design of a Turing Machine relies on the identification of the appropriate language, and the construction of its instruction table will depend on the complexity of the available language; second, the execution of the computation relies on the formulation of the physical operations provided by such instruction table, in terms of reading, writing, and moving on the tape; these operations are physically interpreted and induce corresponding limits. In this sense, the mechanical model of computation by a machine is a further step in the definition of the nature of algorithms in terms of language and implementation, which will be further enhanced by the advent of computer science as a discipline.

The roots of Computer Science in the history of algorithms was pointed out by authors like Donald E. Knuth and Herman Goldstine in the 1970s.[6] Knuth defines computer science as *'the study of algorithms'*[7] and always linked the success of the discipline very explicitly to algorithms:

> I believe that the real reason underlying the fact that Computer Science has become a thriving discipline at essentially all of the world's universities, although it was totally unknown twenty years ago, is not that computers exist in quantity; the real reason is that the algorithmic thinkers among the scientists of the world never before had a home.[8]

The importance of languages for algorithmic representation in computer science is obvious: modes of presentation have been constantly developed and added, from flowcharts and external documentation to formal methods and domain-specific languages.[9] But in this respect, computer science did not essentially diverge from mathematics as an abstract discipline of algorithmic thinking using symbol manipulation. Knuth, while identifying in this form of thinking the common origin of mathematics and computer science, specifies two characteristics of the latter:[10]

---

[5] For an overview of analog algorithms, see Bournez et al. (2018).

[6] See in particular Knuth (1972); Goldstine (1972, 1977). For this reconstruction, see also Bullynck (2015). Goldstine was the liaison officer with the team at the Moore School of Electrical Engineering, University of Pennsylvania, that was building the ENIAC, one of the world's first computers; see Chapter 8. He was responsible for recruiting John von Neumann for the group and, after the war, continued his collaboration with von Neumann, co-authoring various important founding documents in computing.

[7] See (Knuth 1974, p.323)

[8] See (Knuth 1985, p.172).

[9] See e.g. Fant (1993).

[10] See Knuth (1981).

1. algorithms are characterized in terms of complexity classes;
2. algorithms use a dynamic notion of state, syntactically expressed by the assignment operation.

Both these properties are closely related to the issue of implementation. In his foundational *The Art of Computer Programming*, Knuth re-defines algorithms as follows:

Besides merely being a finite set of rules that gives a sequence of operations for solving a specific type of problem, an algorithm has five important features:

1. *Finiteness. An algorithm must always terminate after a finite number of steps. [...]*

2. *Definiteness. Each step of an algorithm must be precisely defined; the actions to be carried out must be rigorously and unambiguously specified for each case. [...]*

3. *Input. An algorithm has zero or more inputs. [...]*

4. *Output. An algorithm has zero or more outputs. [...]*

5. *Effectiveness. An algorithm is also generally expected to be effective, in the sense that its operations must all be sufficiently basic that they can in principle be done exactly and in a finite length of time by someone using pencil and paper.*[11]

This famous definition illustrates how implementability is essential to the nature of algorithm and how this in turn is tied to determining their complexity and the procedural aspect of their execution. In this sense, computing machinery (intended as the infrastructure performing algorithmic procedures) has had an enormous impact on the understanding and explanation of algorithms. To mention some examples, consider the case of the ENIAC, one of the first digital computers: issues like speed and accuracy have influenced the design of its programs and the design of the machine architecture has affected the understanding of its algorithms.[12] Similarly, practical and physical aspects of programming language have been highly relevant in computer simulations and in computational number theory.[13]

Algorithms exist thus at different levels of abstraction: an algorithm can be interpreted as an abstract formulation of rules, but it also requires an implementation in some concrete means of expression, linguistic first and physical afterwards. The engineering foundation of computing, which will be explored in the second part of this volume, will strongly focus on this latter aspect, and this dichotomy will follow us in all the successive incarnations of computing. Hence, while in the context of the mathematical foundation of computing it seems natural to qualify the conceptual identity between the formal set of configurations of a TM and the translation of an algorithm for the

---

[11] See (Knuth 1997, vol.1, pp.4–6).
[12] See Bullynck and De Mol (2010); De Mol et al. (2015).
[13] See respectively Varenne (2013) and Bullynck (2009).

same program; this intuitive association is, nonetheless, not unproblematic. In particular, it requires us to clarify whether a TM should be understood as a formal counterpart of an algorithm, or rather as an implementation. In the former case, the two objects (algorithm and machine) can be seen as denoting the same abstract construct; in the latter case, a conceptual priority of the informal algorithm on the formal machine translation is subsumed. Obviously, by way of the identity suggested by Church's and Turing's Theses, formally it can be shown that a recursive structure can be implemented in a machine model, and that a machine can model a recursive structure. In that sense, both express the same construct. But issues of language and implementation points to a more tight connection between the machine interpretation of algorithms and their physical nature. In other words, it should be clarified whether algorithms are Turing Machines (intended as purely conceptual constructs equivalent to recursive functions), or processes that can be *implemented* by Turing Machines (intended as physical realization of the corresponding abstract constructs). This problem becomes even more clear in the context of modern computing, where the levels of abstraction have multiplied. Consider objects like HEAPSORT, MERGESORT, INSERTIONSORT: nowadays, practitioners hardly think of them in terms of TMs, although every well-trained computer scientist is aware of the formal relation with a corresponding formal model of computation to produce them. It is possible to refer to entirely different objects when using any of those terms.[14] In the following sections, we intend to explore the debate and the possible positions on the nature of algorithms, trying to preserve in our analysis the idea of their layered nature.

6.2 Intuitive Explanation

The quick and brief illustration of the background history of algorithms presented above is meant to introduce their modern treatment, and explain the connection with intuitive and mechanical computability. Recall from Chapter 5 that the notion of effectively computable function given in Definition 35 relies on the idea of a complete set of instructions, given step by step to calculate the value of the intended function on any given input. This aspect can be generalized to provide an intuitive definition of algorithm, a paraphrase of Kunth's more famous list of requirements given in the quote above:

Definition 63 (Algorithm—Intuitive Definition) *An algorithm is a self-contained, step-by-step set of operations to be performed in the process of attaining a goal.*

This generalization abstracts from the input/output relation specific of computability, and refers to a well-defined process aiming at the obtaining of a goal. In this sense, it has been often remarked (also by Knuth himself) that an algorithm is quite like a recipe. But let us spell out this intuition in more detail, and try to identify some essential properties missing in the above intuitive definition.

---

[14] For another analysis of the same kind, see Dean (2016).

First of all, algorithmic processes are *self-contained and clearly expressed* procedures: it is enough that a method exists to establish how to get such a set of operations. This also allows an algorithm to induce more than one set of operations.[15] Secondly, a requirement on *effectiveness* should be added: we want our goal to be reached, not just a definition of the process towards that goal. In this sense, the recipe needs to guarantee that its execution will produce the intended result, possibly without variations, or degrees of perfection. A third issue is the already mentioned crucial aspect of *implementability*: in order to be effective, the provided set of operations needs to be formulated, understood, and executable in a feasible way. This means that the recipe must not only show the steps to be executed (or at least a method to produce such steps), but also how to execute them, and what tools are needed to this aim. Finally, we want to restrict implementability to the most suitable form, which can be called *efficiency*. Hence the recipe should avoid redundant, albeit unproblematic steps that do not contribute to the overall straightforward result.

Collecting these further remarks, we can attempt a more stringent definition:

Definition 64 (Algorithm — Refined Intuitive Definition) *An algorithm is a self-contained method to determine a step-by-step set of operations expressible within a finite amount of space and time, effective and efficient in obtaining a prescribed goal.*

This definition specifies better some of the interesting properties that seem to be expected from well-defined algorithms:

1. an algorithm is *self-contained* if it does not require any external indication for its comprehension and execution;

2. it is *effective* if its execution produces the intended result;

3. it must be composed by a set of operations such that at each step *it is defined what happens at the next one*;

4. its formulation must be such that it *practically* allows it to be understood and formulated;

5. and it is *efficient* if its formulation requires less computational resources than its execution.[16]

The analysis of computable functions and their ability to decide truth for predicates illustrated in the previous chapters can now be linked explicitly to algorithms. To express this relation, we need to reduce the definition of an algorithm given above to the problem of the decidability of a predicate for a set of elements, something we have informally used

---

[15] This aspect is reflected in modern programming languages by the existence of different paradigms to implement algorithms: functional, imperative, probabilistic, and the recent explosion of machine learning methods.

[16] This refined definition can be compared to other informal ones offered in the literature; see e.g. (Rapaport 2012, Appendix), (Rapaport 2018, p.253), and Hill (2016). Note that in this latter contribution the informal definition refers initially to a *task* to be executed and only in its refinement it adds the property of *accomplishing a given purpose*: in our approach we look straightforwardly at the reaching of a *goal* as an essential property.

in Chapter 3, and which can be now more properly defined in terms of the existence of an algorithm:

Definition 65 (Decidability of a Predicate) *A predicate P with domain D is a property of elements of D such that for all $x \in D$ either $P(x)$ or $\neg P(x)$. P is decidable if there exists an algorithm such that for every x it allows to be established whether $P(x)$ is true or not.*

The characteristic function of a set S in Definition 28 establishes for which elements of S a given predicate holds. This function can now be formulated for a predicate P in terms of an algorithm:

Definition 66 (Characteristic Function of a Predicate) *For any predicate P, we can associate a function f with range $\{0, 1\}$ such that, for all $x \in D(P)$, $f(x) = 0$ if $P(x)$ is false and $f(x) = 1$ if $P(x)$ is true. P is decidable if and only if f is computable. If P is decidable, there is an algorithm for defining the set S of elements satisfying P.*

This algorithm is nothing else than a *decision procedure* for the predicate P. For example, consider the decidability of the predicate defining binary addition: given as input two positive numbers m_1 and m_2 presented in binary, the output of the algorithm should be the result $n = m_1 + m_2$ in binary. The algorithm used could be *bitwise addition with carry*, which is succinctly defined by the following description:

$$0 + 0 = 0$$
$$0 + 1 = 1$$
$$1 + 0 = 1$$
$$1 + 1 = 0 \quad \text{and carry 1 to the next more significant bit}$$

Now this algorithm makes the predicate 'being the sum of two binary numbers' a decidable one.

In view of Theorem 11, we know of the existence of at least one uncomputable function (and in fact several more). Accordingly, let us provide a corresponding definition in terms of an algorithm:

Definition 67 (Uncomputable function) *A function f is uncomputable if no algorithm can be devised such that for some predicate P with domain D it can be established for every $x \in D(P)$ if $P(x)$ is true or false, i.e. if $f(x) = 1$ or $f(x) = 0$.*

The *Entscheidungsproblem*, or decision problem given in Proposition 2, was precisely the question about the possibility of providing algorithms for every mathematical problem. Its original formulation was given, thanks to Church's Thesis, in terms of recursion-based procedures. The machine interpretation of computable functions, as computation performed in terms of algorithms as intended by Definition 64, has led us to the formulation of Turing's Thesis that all computable functions are machine computable, and hence the 'algorithmic' interpretation of the decision problem in Proposition 3. The duality expressed by these two formulations is at the basis of the interpretations of algorithms as requiring language and implementation.

To render precisely this distinction, and to understand how the intuitive notion of algorithm introduced in this section is modified by the introduction of its linguistic formulation and its implementation, we explore in the next sections all levels in the layered structure of algorithms.

6.3 Algorithms as Specifications

Algorithms as fully abstract constructs need to be conceptually separated from the procedures they describe. Let us start by an informal presentation of this connection:

Definition 68 (Algorithm as Informal Specification) *An algorithm A is an informal description of a procedure P.*

For example, one might provide the following informal description of a procedure:

> Sort the element of a list by continually splitting it in half. If the list is empty or has one item, it is sorted by definition (the base case). If the list has more than one item, we split the list and (recursively) invoke the current procedure on both halves. Once the two halves are sorted, the fundamental operation is performed of taking two smaller sorted lists and combining them together into a single, sorted, new list.

As we know, this is an algorithm called MERGESORT. An interpretation of algorithm as informal descriptions abstracts entirely from the computation, permitting arbitrary sets of states and transitions between inputs and outputs. Algorithms defined at this level of abstraction guarantee a generality which can then be seen implemented in different programs. The mode of presentation of such abstract construction may vary: for example, in the above algorithm for MERGESORT there is no procedural specification about how to split the list in the first place (to identify where the half is), how to sort each half (which type of sorting) and so on. The aspects that pertain to the language definition and implementation are abstracted from the description. This corresponds to a *denotational reading* of the relation between different possible implementations or programs $P_1, \ldots, P_n$ and their common mathematical construct C: the function associating each P_i with C can be seen as a (possibly partial) input/output function for every possible value given to each P_i; as long as two programs P_i, P_j present the same input/output relation (black box), they denote the same C.

If we are not very strict in requiring that an algorithm expresses a *detailed* procedure for the execution of a task, we come closer to an extensional interpretation of functions, where we do not know how input and output are related, but rather consider the function a sort of black box. While the advantage of considering algorithms at high level of abstraction is their generality, this interpretation is in fact a problem for the very notion of algorithm as a step-by-step, self-contained set of operations as illustrated in Definition 64. Obviously, a lower abstraction level is required.

6.4 Algorithms as Procedures

At a lower level of abstraction, algorithms can be interpreted as procedures, while still being kept conceptually separated from their machine implementations. This interpretation restores the idea that algorithms should account for a set of step-by-step instructions on how to obtain a certain goal. It does so by associating different procedures $P_1, \ldots, P_n$ to the common mathematical construct C, each P_i intended as a relevant mode of execution of C. Technically, this amounts to a *procedural interpretation* of any two different functions for P_i, P_j, here still considered denoting the same constructs C. A definition in this direction requires a reference to the execution of the procedure of interest in some formal language:

Definition 69 (Algorithm as Procedure) *An algorithm A is a formal description in a language L of how to execute a procedure P.*

An example with MERGESORT interpreted procedurally by a standard recursive definition is the following:

Definition 70 (MergeSort Recursive Equation) *For a list*

$$l = \{l_0, l_1, \ldots, l_{n-1}\}$$

the MERGESORT *function is defined as:*

$$sort(l) = \begin{cases} l, \text{ if } |l| \leq 1 \\ merge(sort(h_1(l)), sort(h_2(l))), otherwise \end{cases}$$

where $|l|$ *is the length of l,* h_1, h_2 *are the left and right halves of l and merge is defined as follows:*

$$merge(l, m) = \begin{cases} m, \text{ if } l = \varnothing \\ l, \text{ if } m = \varnothing \\ l_0 * merge(tail(l), m), \text{ if } l_0 \leq m_0 \\ m_0 * merge(l, tail(m)), \text{ if } l_0 \geq m_0. \end{cases}$$

A recent attempt at reviving this approach has been given in terms of a notion of *recursor*.[17] Recall that a partially ordered set (or poset) is a set taken together with a partial order on it. A poset D is complete if every linearly ordered subset $C \subseteq D$ has a least upper bound $supC$. Finally, recall that a mapping $f : D \rightarrow E$ from one poset D to a poset E is continuous if for every chain $C \subseteq D$ and point $y \in D$

$$y = supC \Rightarrow f(y) = supf[C]$$

Then a recursor is defined as follows:

---

[17] See Moschovakis (2001).

Definition 71 (Recursor Moschovakis (2001)) *For any poset X and any complete poset W, a continuous recursor $\alpha : X \rightsquigarrow W$ is a tuple*

$$\alpha = (\alpha_0, \alpha_1, \ldots, \alpha_k)$$

such that for suitable complete posets $D_1, \ldots, D_k$

- *each $a_i : X \times D_1 \times \cdots \times D_k \to D_{i=1\ldots k}$ is a continuous mapping;*
- *the output mapping $\alpha_0 X \times D_1 \times \cdots \times D_k \to W$ is continuous.*

It is crucial that the output mapping has a *fixed least point*, because this guarantees that the recursion terminates. Composition, permutation, currying, conditionals are all admitted on recursors.

Rendering this interpretation of an algorithm means to refer to a (formal) linguistic construct for the execution of the procedure, while at the same time recovering the efficiency aspects of the computation and the specification of its available resources. In particular recursors allow one to define class for algorithms based on the number of relevant operations they require. The input/output relation is now restored in a white-box style. As algorithms are effectively classified by complexity properties related to time, space, and memory, which in turn allow one to establish which problems are feasibly decidable, an obvious requirement is that the procedural description account for those. When considering algorithms as abstract constructors in the more formal approach of general recursors, there are still many different compilation procedures that could be applied for a given recursion schema: for example the

$$merge(sort(h_1(l)), sort(h_2(l)))$$

step could compute $sort(h_1(l))$ first, or $sort(h_2(l))$ first, or compute them in parallel. But they can be classified according to counting operation.

This shows another important aspect: once a sufficiently high level of abstraction is admitted, one is required necessarily to resort to *equivalence classes* of procedures of programs, to account for all the different formulations of the same abstract construct. Hence, at an even lower level of abstraction we are required to indicate both implementability details and equivalence properties among them.

6.5 Algorithms as Abstract Machines

While the approach in terms of recursors allows generality and the preservation of the procedural aspect required by the notion of algorithm, it does not account for some essential aspects associated with implementation. Algorithmic computation is an evolving notion, which has developed from classical sequential algorithms that compute in steps of bounded complexity, through synchronous parallel algorithms, to interactive sequential algorithms, and obviously is completed by analogue algorithms. Some functions implemented by these algorithms are not computable by the Turing model, for example the greatest common divisor $d = gcd(a, b)$ when the lengths are arbitrary and cannot be put on a Turing tape. In general all functions computed by ruler-and-compass algorithms are not Turing computable. Moreover, those that are machine-computable are so in different complexity classes. Recursion does not catch the low-level abstraction of imperative algorithms and software today has very little of declarative specification. Also, complexity aspects related to data structures are not covered by recursion, nor are aspects related to distribution of the processes and non-monotonicity.[18]

These considerations point to the need for interpreting algorithms as formal descriptions that are preserved at an even lower level of abstraction in terms of implementation:

Definition 72 (Algorithm as Implementable Abstract Machine) *An algorithm A is a formal description in a language L specifying the execution of the relevant program P for a given machine M.*

This lower-level definition requires fixing a language L as above, but also the machine M on which P can be executed. For example, MERGESORT can be described linguistically as follows:

```
function mergesort(l)
 var list left, right, result
      if length(l) ≤1
      return l
    else
      var middle = length(l)/2
           for each x in l up to middle
           add x to left
           for each x in l after middle
          add x to right
      left = mergesort(left)
      right = mergesort(right)
      result = merge(left, right)
  return result
```

for any compiler-compatible machine M accepting this pseudo-code linguistic translation.

This approach can be formulated in a more abstract setting. An example of this level of abstraction is given by interpreting sequential algorithms by abstract state machines, a dynamic generalization of the Turing Machine model:[19]

Definition 73 (Sequential Time Algorithm) *A sequential algorithm A is associated with a set $\mathcal{S}(A)$ of states, a subset $\mathcal{I}(A) \subseteq \mathcal{S}(A)$ of initial states, a map $\tau_A : \mathcal{S}(A) \to \mathcal{S}(A)$ of one-step transformations of A.*

---

[18] These observations are at the basis of the analysis proposed in Gurevich (2012).
[19] This model is presented in Gurevich (2000).

With such a state transition function, it is still possible to express the notion of computation as execution:

Definition 74 *A run (or computation) of A is a finite or infinite sequence $X_0, X_1, X_2, \ldots$ where X_0 is an initial state and every $X_{i+1} = \tau_A(X_i)$.*

Under this interpretation of effective computability as implementation, a closer account of a physical computational system becomes possible. This has the advantage of allowing for specific forms of computation, like for sequential algorithms above, or of parallel computations. But the reference to discrete agents imposes constraints on the notion of computation itself. As in the case of the interpretation of algorithms as procedures, this reduction of an algorithm A to its implementation in a given executable language L requires identifying different programs within the same equivalence class, i.e. for different languages. By choosing any given language L (or one of several possible sets of configurations of a TM) implementing the intended algorithm A, one allows several different objects all denoting the same algorithm. For example, equivalence of sequential algorithms is defined by identity of states, initial states, and one-step transformations of the algorithms of interest, inducing equivalence of the corresponding runs: in other words, equivalent algorithms have the same behaviours. It remains to specify more closely the criteria required by such equivalence and to formulate computable relations for such equivalence classes.

6.6 Equivalent Algorithms

The analysis above expresses algorithms through several different levels of abstraction, moving from a more abstract to a more concrete interpretation. An interpretation of algorithms as informal specification is characterized only as an input/out relation, but misses the step-by-step definiteness. A language-based formal interpretation has the property of collecting equivalent families of algorithms, discarding issues of complexity. A machine-based interpretation allows both procedurality and equivalence to be preserved.

Algorithms intended as logical implementable constructs of the corresponding function maintaining specification and implementation (as machine-executable programs) rely on determining equivalence criteria on both the abstraction and the implementation. The identification of such criteria is a long-standing topic in the philosophical debate, to the point of making them criteria for the identification of artefacts in general.[20] The identity relation R is usually characterized by constraints:[21]

Definition 75 (Identity Criteria) *R is an identity relation between two objects x, y iff it satisfies the following constraints:*

---

[20] For the philosophical tradition, see Lowe (1997); Wiggins (2001); Quine (1990). This issue has been investigated at large for the identity of computational artefacts in Angius and Primiero (2018).

[21] See Carrara and Giaretta (2001) and Angius and Primiero (2018).

- **Non-vacuousness**: R must refer to properties that are relevant for defining identity between objects x and y of a given kind K, and such that they not be trivially satisfiable by x and y.

- **Informativeness**: R in the identity criterion is informative with respect to K, in the sense that R should not specify tautological properties.

- **Partial Exclusivity**: R should not appear in the identity criterion for objects of kind different than K and each kind of objects should have its own distinct identity criterion.

- **Minimality**: R should include the smallest set of properties being both necessary and sufficient to determine identity of any two objects of kind K.

- **Non-circularity**: R must not make reference to the identity relation.

- **Non-totality**: $R \subset K \times K$, that is, that not all object pairs in K satisfy the identity relation defined by R.

- **K-maximality**: Given two K objects x and y, for any different relation R' defining identity for K objects, it should hold that $R'(x,y) \rightarrow R(x,y)$ but $\neg(R(x,y) \rightarrow R'(x,y))$.

- **Uniqueness**: for any R', also $\neg(R'(x,y) \rightarrow R(x,y))$ holds.

- **Equivalence**. R is required to be reflexive, symmetric, and transitive.

- **Congruence**. If $R(x,y)$, then all properites satisfied by x are all and only the properties satisfied by y.

The class of programs identical for a given algorithm must then be an equivalence relation:[22]

Definition 76 (Equivalence Relation over Algorithms) *A set of algorithms $A_1, \ldots, A_n$ satisfy an equivalence relation $\approx$ over the class of programs $P_1, \ldots, P_n$ from the different languages $L_1, \ldots, L_n$, executable on different machines $M_1, \ldots, M_n$ if and only if for each (A_i, A_j) in the same class $\approx$ it holds an identity relation $R(A_i, A_j)$ as by Definition 75.*

To explain the equivalence relation $\approx$, one wants to be able to say when the algorithm A expressed by a program P_i and implemented by a machine M_i is equal to an algorithm expressed by a program P_j and implemented by a machine M_j in terms of a relation $P_i \approx P_j$ (or $M_i \approx M_j$). A standard reference for this task is Milner's algebraic definition known as *simulation*,[23] consisting of a uniform transition-preserving mapping between the states comprising the computations induced by two machines M_i, M_j from different models; to establish a precise sense in which two programs realize the same algorithm is sufficient (but not necessary) to show that the programs compute the same states in the same sequence. For the case of sequential algorithms, one can use the association of the

---

[22] Which of these constraints are satisfied by which relation between algorithms can lead to an understanding identity on algorithms as weaker relations. For the analysis at the level of formal specifications expressed as state transition systems and for the weakenings of this notion required by different relations of copy, see again Angius and Primiero (2018). The analysis above shows instead that such a relation can be accounted for both at the level of procedures and at the level of machines.

[23] See Milner (1971).

algorithm A to an abstract state machine computing it $ASM(A) := \{\mathcal{S}(A), \mathcal{I}(A), \tau_A\}$ as by Definition 73.[24] The equivalence of ASMs is formulated by the standard refinement method as follows:[25]

Definition 77 (ASM Refinement) *Fix any notion $\equiv$ of equivalence over (initial and final) states. An ASM M^* is called a correct refinement of an ASM M if and only if for each M^*-run $S_0^*, S_1^*, \ldots$ there is an M-run $S_0, S_1, \ldots$ and sequences [of states] $i_0 < i_1 < \ldots, j_0 < j_1 < \ldots$ such that $i_0 = j_0 = 0$ and $S_{i_k} \equiv S_{j_k}^*$ for each k and either*

- *both runs terminate and their final states are the last pair of equivalent states; or*
- *both runs and both sequences $i_0 < i_1 < \ldots, j_0 < j_1 < \ldots$ are infinite.*

M^* is called a complete refinement of M if and only if M is a correct refinement of M^*.

A one-way relation between two such machines is expressed by simulation:[26]

Definition 78 (Simulation) *The M^*-run $S_0^*, S_1^*, \ldots$ is said to simulate the M-run $S_0, S_1, \ldots$ if and only if M^* is a correct refinement of M.*

Correct refinements with respect to terminating runs preserve so-called partially correct implementations; total correctness adds that every infinite refined run admits an infinite abstract run with an equivalent initial state. The relation can be generalized to a bisimulation:[27]

Definition 79 (Bisimulation) *The M^*-run $S_0^*, S_1^*, \ldots$ and an M-run $S_0, S_1, \ldots$ are in a bisimulation relation if and only M^* is a correct refinement of M and vice versa (i.e. M^* is a complete refinement of M).*

Note that this refinement relation is the one used to formulate the relation between an abstract state machine and each of its implementations. Then one can define equivalence of the algorithms executed by programs through bisimulation on abstract state machines:

Definition 80 (Algorithm Equivalence) *A machine M_i running a program P_i in language L_i executes the same algorithm A of a machine M_j running program P_j in language L_j if and only if given a ASM interpreting M_i and given a ASM' interpreting M_j, then M_i is a correct and complete refinement of M_j.*

---

[24] The problem of automatic verification of ASM-programs, i.e. to decide given a program P for an ASM and a correctness property ϕ whether this property is satisfied by any execution of P on any input i is known to be decidable, see Spielmann (1999). The more general model-checking problem (where the property ϕ is not constrained to be a correctness property) and the general verification problem (where the validity of ϕ must be guaranteed for every input) are in general undecidable, while restricted versions can be proven decidable, see Spielmann (2000). In the following we are interested in the more abstract problem of formally deciding equivalence of ASMs based on the identity criteria above.

[25] See Börger (2003).

[26] An equivalent relation for state transition systems is used to define the notion of inexact copy in Angius and Primiero (2018).

[27] An equivalent relation for state transition systems is used to define the notion of exact copy in Angius and Primiero (2018).

The equivalence of distinct algorithms within the same equivalence class can be correspondingly expressed by simulation:

Definition 81 (Algorithm Class Equivalence) *A machine M_i running a program P_i in language L_i executes an algorithm A in the same class of a machine M_j running program P_j in language L_j if and only if given a ASM interpreting M_i and given a ASM' interpreting M_j, then M_i is a correct refinement of M_j.*

According to this view, algorithms are still abstract objects, but are expressed in terms of corresponding machines executing them, given their translation in an appropriate language. This interpretation of the notion of algorithm is interesting also for the understanding of computing as a mathematical discipline in relation to its concrete aspects, as they surface when considering effective computability by machines. A purely abstract, non-formal definition of algorithm has the advantage of generality, and thus it allows for a larger scope; but it makes it impossible to account for aspects related to implementation, its procedural nature, and the rendering of different equivalent programs and their complexity properties. This leads us, in the first place, to reject the identity between algorithms and specifications.[28] The more formal treatment guaranteed by an analysis in terms of recursion has the limitation of not covering aspects of algorithms proper of machine implementable procedures. More generally, this shows a first weakening of the purely mathematical way of understanding computing: despite the identity of the different models of computation, once the mechanical notion is introduced, there are properties (structure, boundedness, locality, parallelism) that do not easily transfer back to other formally equivalent counterparts. In this chapter we have insisted on algorithms as objects with a layered ontology, as abstract implementable logical constructs, in order to be able to maintain both their abstract nature and their physical nature. An effect of requiring algorithms to be understood in this sense is that an appropriate identity relation on them becomes essential. While this might seem at first an additional conceptual and formal burden for a notion that has an intuitive and immediate appeal in practical terms, it has the advantage of providing the essential machinery to preserve the generality shared by different instances of the same algorithm. It allows one to decide when two programs (procedures, functions) are instances of the same algorithm.

More essential is the need to decide when a program (procedure, function) is the correct one for the problem at hand: this means selecting the appropriate equivalence class of programs. This leads us to the issue of *program correctness* to be explored in Chapter 7. Also in this case, and to conclude the first part of this volume, we will consider the approach to correctness grounded on the mathematical roots of computing, with programs understood as counterparts of proofs. As for the case of algorithms, this debate will lead us to the dicotomy between the formal and the physical aspects of computing.

---

[28] The latter is a notion that requires its own analysis and which will be considered properly in Chapter 12.

Exercises

Exercise 56 *Define algorithms in informal terms.*

Exercise 57 *Which characteristics define algorithms in computer science compared to mathematics?*

Exercise 58 *Formulate the intuitive definition of algorithms.*

Exercise 59 *Define decidability of a predicate in terms of algorithms.*

Exercise 60 *Define uncomputability of a predicate in terms of algorithms.*

Exercise 61 *Define algorithms as procedures.*

Exercise 62 *Define algorithms as abstract machines.*

Exercise 63 *Define bisimulation on ASMs.*

Exercise 64 *Define equivalent algorithms through bisimulation.*

Exercise 65 *Define simulation on ASMs.*

Exercise 66 *Define class equivalence on algorithms through simulation.*

7 Computing as a Mathematical Discipline

Summary

The analysis of algorithms in Chapter 6 has shown the dichotomy between abstraction and implementation. This issue is closely linked to the problem of correctness: conditions of refinements express how to establish correctness of implementations with respect to their abstract machine. In this chapter we explore more closely the debates concerning correctness, we illustrate the principles of formal verification for programs and the philosophical critiques to it. We finally close the first part of this volume by formulating principles of formal computational validity.

7.1 Proofs as Programs

The interpretation of computation given by the model of a machine defined by Turing and the different interpretations of the notion of algorithm surveyed in Chapter 6 have highlighted the deep connection existing between the logical notion of provability and the one of mechanical computation.

Proof was at the basis of the formal approaches to truth underlying Hilbert's programme, and it can be traced back to Frege's *Begriffsschrift*, where a system of sound and complete inference rules (still referred often to as *Frege's System* or *Hilbert's System*) was defined. In its minimal propositional formulation, such a system expresses an implicative fragment of a logical language with an additional axiom for identity and the basic Modus Ponens rule; see Figure 7.1.[1]

---

[1] Modern versions of Frege's Systems are usually completed by an appropriate axiom to introduce negation: $\vdash (\neg A \to \neg B) \to (B \to A)$.

On the Foundations of Computing. Giuseppe Primiero, Oxford University Press (2020). © Giuseppe Primiero.
DOI: 10.1093/oso/9780198835646.001.0001

$$\vdash A \to A$$

$$\vdash A \to (B \to A)$$

$$\vdash (A \to (B \to C)) \to ((A \to B) \to (A \to C))$$

$$\frac{\vdash A \quad \vdash A \to B}{\vdash B}$$

Figure 7.1 Propositional Frege's System

$$\frac{}{A \vdash A}Id$$

$$\frac{\Gamma \vdash A \qquad \Delta \vdash B}{\Gamma, \Delta \vdash A \wedge B} \wedge\text{--}intro$$

$$\frac{\Gamma \vdash A \wedge B}{\Gamma \vdash A} \wedge\text{--}elim(left) \qquad \frac{\Gamma \vdash A \wedge B}{\Gamma \vdash B} \wedge\text{--}elim(right)$$

$$\frac{\Gamma, A \vdash B}{\Gamma \vdash A \supset B} \supset\text{--}intro \qquad \frac{\Gamma \vdash A \qquad \Delta \vdash A \supset B}{\Gamma, \Delta \vdash B} \supset\text{--}elim$$

$$\frac{\Gamma \vdash A}{\Gamma \vdash A \vee B} \vee\text{--}intro(left) \qquad \frac{\Gamma \vdash B}{\Gamma \vdash A \vee B} \vee\text{--}intro(right)$$

$$\frac{\Gamma \vdash A \vee B \quad \Delta \vdash A \supset C \quad \Delta \vdash B \supset C}{\Gamma, \Delta \vdash C} \vee\text{--}elim$$

Figure 7.2 Gentzen's Natural Deduction Calculus

Gerhard Gentzen offered a translation of these principles in his natural deduction system of logic.[2] The context in which his system was developed was still Hilbert's Programme, and in particular the requirement of consistency for mathematics. Assuming the translation of mathematical statements to sentences of a formal language, Gentzen developed rules for all logical operators which could occur in such sentences: the rules were meant to offer the semantics of each connective, by expressing how to *introduce* and how to *eliminate* a formula containing that operator under a fixed set of assumptions (as commonly expressed by Greek capital letters); see Figure 7.2. The introduction rules tell how to define the connective, the elimination rule how to use it, by referring only to the subformulas linked by it: for example, the formula $A \wedge B$ is introduced by having available the subformulas A, B and it is eliminated by obtaining both. With this system, consistency of a system is proven by reduction (or *normalization*) of any of its proofs to one which only contains subformulas of the theorem (the so-called Normalization Theorem).

---

[2] See Gentzen (1935).

Recall from Chapter 4.4 how Church introduced the λ-calculus by reducing every expression to an abstraction function of any number of arguments, reducible to applied terms, including the possibility of taking functions themselves as arguments of other functions. The problematic aspect of circularity (see again Chapter 2) induced by the unconstrained applicability of functions to functions was resolved by Church in his formulation of the *typed* version of the λ-calculus.[3] Now each formula A is replaced by one decorated with an appropriate term $x : A$ so that type annotations are given in abstraction, and application is constrained. Curry, a pupil of Church, provided an interpretation in which an intuitive correspondence is established between a function $f : A \rightarrow B$ with an input of type A and an output B with a corresponding implication $A \supset B$. In his work, a system of axioms for the implicational fragment of logic (i.e. the same object of Frege's and Gentzen's systems) is analysed in terms of:

statements of the form (using E. H. Moore's terminology) 'f is a function on X to Y'; or (for functions of several variables) 'f is a function on $X_1 X_2 \ldots X_m$ to Y'.[4]

effectively showing the possibility of reducing the semantics of implications to functions. Accordingly, for any computable function there is a provable proposition of that form, and vice versa. The principle was then later independently reformulated by William Alvin Howard and Nicolas de Bruijn in the 1960s and 1970s respectively.[5] Howard extended Curry's intuition to the other logical connectives, making use of results by Prawitz:[6] now every natural deduction pair of rules can be reformulated in the typed λ-calculus; see Figure 7.3.[7] The typing operation, while making sure that applications are constrained to appropriate predicates and that self-referential terms cannot be expressed, also constrains the expressive power of the calculus: the untyped version is Turing-complete, i.e. it can express any function formulated by any Turing machine, while the typed version cannot express recursion.[8]

This formal correspondence between existential proofs and executable programs became a more precise way of expressing the identity between mechanical computability and provability by recursive means expressed by the Church-Turing Theses. Following the line of works here mentioned, the identity became known as the *Curry-Howard-de Bruijn Correspondence*. The most abstract formulation of the principle refers to the identification of propositions as categories of expressions:

---

[3] See Church (1934).

[4] See (Curry, 1934, p.585).

[5] See Howard (1980) and de Bruijn (1970, 1980). For another recent overview of this principle and its conceptual and historical roots, see Wadler (2015). For in-depth treatment of the formal aspects related to the propositions as types identities, see Girard et al. (1989), Thompson (1991), Sørensen and Urzyczyn (2006).

[6] See Howard (1980) and Prawitz (1965).

[7] In the elimination rules for conjunction, π_1, π_2 stand respectively for the first and second projection of the term t, which by its introduction we know must be a pair. In the rules for disjunction, the operators *inl, inr* stand respectively for injection left and injection right of the term for the disjunction $A \vee B$.

[8] Recursion can be recovered by adding appropriate terms but losing termination.

$$\frac{}{x : A \vdash x : A}Id$$

$$\frac{\Gamma \vdash t : A \qquad \Delta \vdash u : B}{\Gamma, \Delta \vdash \langle t, u \rangle : A \wedge B} \wedge{-}intro$$

$$\frac{\Gamma \vdash t : A \wedge B}{\Gamma \vdash \pi_1(t) : A} \wedge{-}elim(left) \qquad \frac{\Gamma \vdash t : A \wedge B}{\pi_2(t) : B} \wedge{-}elim(right)$$

$$\frac{\Gamma, x : A \vdash t : B}{\Gamma \vdash \lambda x.t : A \supset B} \supset{-}intro \qquad \frac{\Gamma \vdash t : A \qquad \Delta \vdash u : A \supset B}{\Gamma, \Delta \vdash t(u) : B} \supset{-}elim$$

$$\frac{\Gamma \vdash t : A}{inl(t) : A \vee B} \vee{-}intro(left) \qquad \frac{\Gamma \vdash u : B}{inr(u) : A \vee B} \vee{-}intro(right)$$

$$\frac{\Gamma \vdash t : A \vee B \quad \Gamma, x : A \vdash u : C \quad \Gamma, y : B \vdash v : C}{\Gamma \vdash \lambda x.t(u) \mid \lambda y.t(v) : C} \vee{-}elim$$

Figure 7.3 Typed λ-calculus

Principle 3 (Curry-Howard-de Bruijn Correspondence) *Propositions are Types.*

This principle, in turn, expresses an identity between the categories proper of logic and those of programming by stating that logical formulas can be interpreted according to the type of function satisfying them, assuming the latter are restricted to computable ones.

Howard also offered new types for the quantifiers of first-order logic: the corresponding functional constructions, nowadays known as dependent types, express the value of a predicate $B(x)$ when x is assumed to satisfy another predicate A. This interpretation of function types reduces to the semantics of logical connectives under the intuitionistic reading already mentioned in Chapter 2. Recall that for the intuitionists the problem of avoiding an infinitely proceeding sequence in mathematics meant to admit only sequences which determine successively all of their predicates through well-defined rules. For the general semantics of logical sentences, this meant that the truth of a proposition could only be guaranteed based on its proof. Accordingly, the Law of Excluded Middle $A \vee \neg A$, for every proposition A was intuitionistically invalid. The BHK semantics, from the initials of Brouwer-Heyting-Kolmogorov, is thus given in terms of proof objects, and truth reduced to their construction (see again Definition 16). All these ideas were conducive to the development of an Intuitionistic Type Theory by Martin-Löf.[9] Propositions identified with the set of their proofs induce a correspondence to function types, Figure 7.4 and Figure 7.5:

---

[9] See Martin-Löf (1984). For introductions to Martin-Löf's Type Theory, see Nordström et al. (1990), Primiero (2006), Grantsröm (2011).

FUNCTION TYPE	PROPOSITION
$A \rightarrow B$	$A \supset B$
$A \times B$	$A \wedge B$
$A + B$	$A \vee B$
$\lambda(b) : \Pi(A,B)$	$(\forall x : A)B(x)$
$(a,b) : \Sigma(A,B)$	$(\exists x : A)B(x)$

Figure 7.4 Correspondence between function types and propositions

$$\frac{a : A \qquad b : B}{(a,b) : A \wedge B} \qquad \frac{(a,b) : A \wedge B}{\pi_1(a,b) : A \text{ and } \pi_2(a,b) : B}$$

$$\frac{x : A \vdash b : B}{\lambda((x)b) : A \rightarrow B} \qquad \frac{\lambda((x)b) : A \rightarrow B \qquad a : A}{b(a) : B(a)}$$

$$\frac{a : A}{l(a) : A \vee B} \qquad \frac{b : B}{r(b) : A \vee B}$$

$$\frac{x : A \vee B \qquad \lambda((x)c) : A \rightarrow C \qquad \lambda((y)c) : B \rightarrow C}{c(a) : C(a) \text{ or } c(b) : C(b)}$$

$$\frac{x : A \vdash b : B(x)}{\lambda((x)b) : (\forall x : A)B(x)}$$

$$\frac{\lambda((x)b) : (\forall x : A)B(x) \qquad a : A}{b[x/a] : B}$$

$$\frac{a : A \qquad b : B(a)}{(a,b) : (\exists x : A)B(x)}$$

$$\frac{z : (\exists x : A)B(x) \qquad x : A, y : B(x) \qquad c(x,y) : C(x,y)}{(z,(x,y)c(x,y)) : C(z)}$$

Figure 7.5 BHK Semantics

- implication from a proposition A to B corresponds to a function such that given a proof of A yields a proof of B;
- conjuction corresponds to the Cartesian product, i.e. a proof of A and a proof of B;
- disjunction corresponds to union, i.e. a proof of A or a proof of B;
- the universal quantifier corresponds to a function which given any proof a of A yields a proof of $B(a)$, which in turn is the Cartesian product of the family of sets of proofs a of A and $B(a)$, also known as the Π type;

- the existential quantifier corresponds to a pair whose first component is a proof a of A and the second component a proof b of $B(a)$, which in turn is the disjoint union of the family of sets of proofs a of A and $B(a)$, also known as the Σ type.

A further identity allowed by the intuitionistic interpretation of logical constants, and in particular by Kolmgorov's reading of it, is that the proposition A corresponds to a task A and a proof a of A corresponds to a program for the task whose specification is expressed by A.

Definition 82 (Kolmogorov's Truth) *Task A is solved if and only if there exists a program a for the specification of A.*

Hence, the language of a formal (constructive, existential) proof can be used as a programming language.[10] This leads to the properly computational version of the correspondence, stating that for any proof of a proposition there is a program of the corresponding type:

Principle 4 (Curry-Howard-de Bruijn Correspondence (Computational)) *Proofs are Programs.*

The intuition behind such correspondence is that the computational content of a proof can be given in terms of the (recursive, or so equivalently defined) procedures required to verify a theorem that expresses the valid output of a program. Two elements are thus essential to express the logical structure of programs underlying their identity with proofs: types and recursion. This is spelled out in Figure 7.6.[11]

In terms of programming constructs, iteration is available through the use of selectors on inductively defined sets, and termination is guaranteed by allowing primitive recursion, but general recursion is not available in the basic theory, although it can be added through the definition of appropriate higher-order functions.[12] Conditions in the recursion corresponds to constraints on the input that allow valid behaviour of the programs, while nested implications correspond to higher-order functions.

The identity between recursion and `while` loops was an indirect result of Kleene's Normal Form Theorem given in Definition 44, Turing's formulation given in Theorem 4, and what has become an established result in computer science:[13]

Theorem 18 (Harel (1980)) *Every flowchart is equivalent to a* `while`*-program with an occurrence of* `while-do`*, provided additional variables are allowed.*

This identity became a crucial issue in the debate among different approaches to program design during the 1950s and the 1960s of the twentieth century, when the mathematical foundation of computing started to influence the creation of programming

---

[10] See Constable (1971, 1983).

[11] See (Martin-Löf, 1982, p.168).

[12] See e.g. Paulson (1986).

[13] This result is usually attributed to Böhm and Jacopini (1966). For an actual in depth reconstruction of the results building up towards this, see Harel (1980).

Programming	Mathematics
program, procedure, algorithm	function
input	argument
output, result	value
$x := e$ (assignment)	$x = e$
$S_1; S_2$	composition of functions
if B then S_1 else S_2	definition by cases
while B do S	definition by recursion
data structure	element, object
data type	set, type
value of data type	element of a set, object of a type
$a : A$	$a \in A$
integer	Z
real	R
boolean	$\{0, 1\}$
$(c_1, \ldots, c_n)$	$\{c_1, \ldots, c_n\}$
array $[I]$ of T	$T^I, I \to T$
record $s_1 : T_1; s_2 : T_2$ end	$T_1 \times T_2$
record case $s : (c_1, c_2)$ of $c_1 : (s_1 : T_1); c_2 : (s_2 : T_2)$ end	$T_1 + T_2$
set of T	$\{0, 1\}^T, T \to \{0, 1\}$

Figure 7.6 Key notions of programming with mathematical counterparts

languages inspired by principles of universality and machine independence.[14] This new approach to computing was concentrating on how to realize the mathematical principles of computation in physical systems. When computing machinery was quickly being developed, some of the main actors involved would look back at its mathematical foundation and recognize it as its most fundamental model to interpret the processes embedded in physical computing. This meant, in those and the successive decades, to define the relationship between computing and mathematics. One of the earliest moments when the relation between computer science and mathematics was put under scrutiny is the *Automatic Programming for Digital Computers* Conference held in Washington D.C. in 1954. On this occasion, contributors were focusing on aspects concerning the construction and identification of general properties of systems.[15] At this stage, the problem of standards for programming languages was approached often from the Church-Turing Thesis result, i.e. the assertion that the sequence of sentences in code of instructions can be seen either as the recursive definition of its output or as the set of specifications of a special purpose machine designed for that output, and that the Universal Turing Machine can emulate

---

[14] Many of the events, contributions, and debates recounted in this chapter are investigated at length both in Daylight (2012) and Tedre (2015). The present overview relies on those important historical reconstructions.

[15] See Gorn (1954); Brown and Carr (1954).

any special machine.[16] The issue of defining general properties for languages reflects the bottom-up process required for algorithms in the previous Chapter 5. Between 1958 and 1960, a collaboration was initiated between the *Association for Computing Machinery* and the *Gesellschaft für Angewandte Mathematik und Mechanik* for the definition of a universal algorithmic language, first denoted IAL, later ALGOL58, and finally ALGOL60. In what became known as the 'ALGOL Effort', the idea of universality, machine independence, and the debate concerning the introduction in the language of recursive procedure were all central.[17]

7.2 Program Correctness

The idea that essential properties of programs for physical machines could be established through the analogy with mathematical concepts became a major research program from the early days of computing. Of all the properties of programs that would be object of mathematical analysis, correctness is certainly the most important and most studied one. Formal methods enthusiasts like Donald Knuth, Edsger Wybe Dijkstra, and Sir Charles Antony Richard Hoare supported a purely mathematical perspective on programs, with correctness intended as a property to be established relative to specification through mathematical proof in the spirit of the Curry-Howard-de Bruijn correspondence.[18]

Let us start by offering an informal definition of the problem:

Definition 83 (Program Correctness Problem) *Given any program P, is it possible to prove whether P when executed on the appropriate input i returns the intended output o, or in other words whether it satisfies its intended function?*

This problem expresses the principle that a formally correct program is one whose correctness can be proved mathematically. Proving logically that a program is correct with respect to a given specification means to show that the implementation correctly satisfies the specification. For specification one understands a system where a set of initial states are transformed into a final state, not considering how the transformation takes place, only the relation between states.

The early tradition

The earliest steps in program verification can be traced back to the works of Goldstine and von Neumann, Turing, and Curry.[19] Their research focused on the identification of tractable notations to describe and check the behaviour of programs: flowcharts for

---

[16] See in particular Gorn (1957).

[17] The history of ALGOL is extensive. For the aspects of interest to our analysis, see Daylight (2011); Alberts and Daylight (2014).

[18] For a full recount of the debate, see also (Tedre, 2015, ch.6).

[19] See Goldstine and von Neumann (1947), Turing (1949), and Curry (1949).

Goldstine and von Neumann; diagrams and annotations in Turing's paper.[20] But the work that is usually considered to have started the mathematical tradition on proving program correctness is due to Floyd, who started this analysis by providing an association of steps in the control flow of a program with logically holding propositions.[21] Such associations are prevented to be arbitrary: a relation of logical validity is inductively defined over a proposition at each execution of a program step. The semantic definition is syntactically based: it specifies which linguistic constructs of the language express valid commands, and which conditions hold for each given command. Starting from a flowchart model of the program in which each step is a command, an interpretation corresponds to a mapping of logical sentences to each transition from step to step (called the entry and exit of the command). A verification of the flowchart of the program is a proof that for every command expressed in it, if there is an entrance such that a given sentence is true, then there must be an exit such that another sentence is true. A semantic definition of a particular set of command types corresponds to formulating verification conditions on the antecedents and the consequent of any command of each of these types. An example of such a flowchart for the program to compute the sum of numbers from 1 up to a given integer is offered in the original paper; see Figure 7.7.

The holding of given relations at the beginning and end of program execution is interpreted to prove the correctness and (separately) termination of the program in a purely mathematical way, through translation to a particular set of axioms and inference rules. The issue of correctness and termination were already made distinct much earlier by Turing:

> The programmer should make assertions about the various states that the machine can reach. [...] The checker has to verify that [these assertions] agree with the claims that are made for the routine as a whole. [...] Finally the checker has to verify that the process comes to an end.[22]

A verified program entered at some step that makes a corresponding sentence true must be followed at each subsequent time by a step which makes a corresponding sentence true. If the program halts, the step for the corresponding output state must become true at some point. In the words of Floyd:

> we have not proved that an exit will ever be reached; the methods so far described offer no security against nonterminating loops. To some extent, this is intrinsic; such a program as, for example, a mechanical proof procedure, designed to recognize the

---

[20] The reconstruction of the missing influence of these works on the later better-known contributions by Floyd and Hoare is investigated in Jones (2003) and Jones (2017).

[21] See Floyd (1967).

[22] (Turing, 1949, p.68). There is no indication that Turing's work influenced Floyd's, who gives credits for these proofs to ideas of Perlis and Gorn Gorn (1959). Thanks to Simone Martini and Edgar Daylight for help in reconstructing this order of references with respect to the separation between correctness and termination, which would eventually lead to the formulation of the notion of partial correctness.

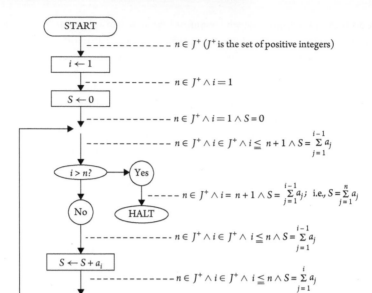

Figure 7.7 Flowchart from Floyd (1967)

elements of a recursively enumerable but not recursive set, cannot be guaranteed to terminate without a fundamental loss of power. Most correct programs, however, can be proved to terminate.[23]

This passage reflects in the reference to non-termination the general undecidability of validity, while stressing that any given derivation can be checked for correctness. The basic idea underlying the approach is thus that properties of such programs in a given language are independent of their physical instantiation and of processors for that language, and can be reduced to establishing standards of rigour for proofs about programs in the given language. This is the principle of machine independence which was guiding the design of languages like ALGOL60.

After Floyd, it was Hoare's work to support a strongly mathematical view of program correctness, what is known as *axiomatic semantics*. From the very start, Hoare is quite explicit in his assumption:

Computer programming is an exact science in that all the properties of a program and all the consequences of executing it in any given environment can, in principle, be found out from the text of the program itself by means of purely deductive reasoning.[24]

[23] (Floyd, 1967, p.30).
[24] (Hoare, 1969, p.576).

This assumption justifies the development of a logic to reason about programs, i.e. to prove properties of programs. The system includes axioms for arithmetics, and expressions of the language refer to values of relevant variables before and after the execution of the program. Any such expression is called an *assertion* and it corresponds to a predicate describing the state of a program at any point of its execution. The starting point of the analysis is thus to determine the valid assignments of variables *before* program execution, i.e. assertion for initial states of the program, or what have become known as *preconditions*. The problem of program correctness is then reinterpreted as the problem of checking that for any program with a given set of preconditions, its execution leads to the expected set of variable assignments or program states *after* the program operations have been performed, or what have become known as *postconditions*. This is interpreted in terms of triples consisting of the precondition of execution P, the program Q, and the postcondition R[25]

$$P\{Q\}R$$

to be read as follows:[26]

> If the assertion P is true before initiation of the program Q, then the assertion R will be true on its completion.

Here, again, correctness and termination are kept separated by the proviso that the program completes its run. An example can be given as follows: consider the empty precondition $P := true$ and a program Q to compute the *max* function between two integers, as follows:

```
int Max(int a, int b){
    int m;
    if (a >= b)
        m=a;
    else
        m=b;
    return m
    }
```

Then its postcondition R is expressed by $(m = max(a,b))$. Hoare provides axioms and rules for such expressions of program execution; see Figure 7.8. In the *Assignment* axiom, P' results from assigning values to variables in the starting state P. In general, the program Q changes the values of the variables it manipulates. *Sequencing* (or *Composition*) says that if starting in a state satisfying P, program Q_1 leads to a state satisfying R and starting in a state satisfying R, program Q_2 leads to a state satisfying S, then starting in state

---

[25] We maintain here the original notation from Hoare (1969): note that here P is a precondition and not a program (as throughout this volume), while the latter is denoted by Q.

[26] See (Hoare, 1969, p.577).

$$\frac{\overline{\phantom{P\{x := v\}P'}}}{P\{x := v\}P'}\textit{Assignment}$$

$$\frac{P\{Q_1\}R \qquad R\{Q_2\}S}{P\{Q_1; Q_2\}S}\textit{Sequencing}$$

$$\frac{P \vdash P' \qquad P'\{Q\}R' \qquad R' \vdash R}{P\{Q\}R}\textit{Consequence}$$

$$\frac{P \wedge R\{Q\}P}{P\{\textit{while } R \textit{ do } Q\}\neg R \wedge P}\textit{Iteration}$$

Figure 7.8 Rules of Floyd-Hoare Logic

satisfying P, the sequential composition of programs Q_1 and Q_2 leads to a state satisfying S. *Consequence* states that if a state satisfying P implies one satisfying P', and from this program Q leads to state satisfying R' and a state satisfying R' implies one satisfying R, then starting from a state satisfying P, execution of program Q leads to a state satisfying R. In the rule for *Iteration*, the invariant P expresses a relationship between values preserved by any execution of Q: the premise states that given P and R are true before we execute Q, if after its execution P is still true, then a program can be designed that expresses execution of Q while R holds and this will have P as invariant.[27]

Proving program correctness means in this setting to describe rigorously the intention of the user as assertions about the variables at the end of the program execution (its postconditions) and checking that these can be in fact reached when the program is executed given its preconditions.

Definition 84 (Hoare Correctness) *A program is correct with respect to a pre- and postcondition specification if the execution of the program on a machine whose initial state satisfies the precondition will result in the machine terminating in a state which satisfies the postcondition.*

Hoare aimed at solving in the first instance three of the main problems of programming:

1. specification satisfaction;
2. program documentation;
3. machine independence.

The axiomatic method is deemed of high value to these aims because it is *general*, implementations can be designed so that they should be required to satisfy the axioms, and it is *universally applicable*, so that aspects of the program—like data-types—can be

---

[27] In this set of rules we reformulate those originally given in Hoare (1969). An extension of this set can be given by adding a rule for *if-else* statements that do not need invariants; see (Huth and Ryan, 2004, ch.4).

left underspecified. The resulting system, sometimes today referred to as Floyd-Hoare logic, is aimed at proving program correctness, assuming that implementation of the programming language conforms to the axioms and rules used in the proof, and the latter are considered explicitly in the absence of side effects of the evaluation of expressions and conditions,[28] a very important abstraction on the actual running and execution of a written program.

In the following decade, a crucial contribution to this debate was given by Dijkstra with his *predicate transformer semantics*: its formulation allows both for a recurrence relation that accounts for general recursion and a functional interpretation of Hoare triples. In this model it is possible to define a notion of correctness which also guarantees termination. This description brings to a further clarification of the notion of states, establishing for each program which terminates and satisfies a postcondition state, the unique precondition state required to realize the appropriate Hoare triple:[29]

Definition 85 (Weakest precondition) *The weakest precondition, denoted $wp(P, R)$ is the condition that characterizes the set of all initial states such that execution of program P will result with certainty in a final state satisfying a given postcondition R.*

The semantics of the program P is given by determining for any post-condition R the corresponding weakest precondition $wp(P, R)$. If the initial state does not satisfy a weakest precondition, P may not establish the truth of R in its final state, or P may fail to reach a final state at all, because it enters an infinite loop or it gets stuck.[30] Hence, while Hoare logic admits for partial correctness (i.e. in the absence of termination), the logic of weakest preconditions allows one to distinguish it from total correctness, reformulated as follows in terms of weakest preconditions:

Definition 86 (Total Program Correctness) *For a given specification*

$$(pre(P), R)$$

a program P is totally correct with respect to its specification iff

$$pre(P) \rightarrow wp(P, R)$$

An early formulation of a weaker notion of *partial correctness* was provided by Manna.[31] Starting from an algorithm for transforming a program P into

- a first-order formula W_P such that it is satisfiable if and only if either P is correct or P does not terminate, and
- a first-order formula $\tilde{W}_P$ such that it is unsatisfiable if and only if P is correct,

then

---

[28] See (Hoare, 1969, pp.578–9).
[29] (Dijkstra, 1975, ch.3).
[30] See (Dijkstra, 1975, p.17).
[31] See Manna (1969).

- the program P is correct with respect to an input predicate $\phi(x)$ and an output predicate $\psi(x,z)$ if for every ξ such that $\phi(\xi) = \texttt{true}$, $P(\xi)$ is defined and $\psi(\xi, P(\xi)) = \texttt{true}$;
- and P is partially correct if for every ξ such that $\phi(\xi) = \texttt{true}$, if $P(\xi)$ is defined then $\psi(\xi, P(\xi)) = \texttt{true}$.

It can be proven that P is partially correct if and only if $W_P[\phi, \psi]$ is satisfiable and P is correct if and only if $\tilde{W}_P[\phi, \psi]$ is unsatisfiable.[32] Informally, we can use the following definition:[33]

Definition 87 (Partial Program Correctness (Informal)) *Proving a program partially correct means proving only that an acceptable answer will be produced under the additional assumption that the computational process terminates.*

This, again, can be translated more formally in terms of the notion of weakest precondition:

Definition 88 (Partial Program Correctness (Formal)) *A program P is partially correct if it can be proved that*

$$pre(P) \wedge wp(P, true) \rightarrow wp(P, R)$$

where $wp(P, true)$ is the weakest precondition ensuring termination in any state.

The combination of partial correctness and termination can be offered as follows:

Definition 89 (Total Correctness—Alternative Definition) *A program is correct with respect to a pre- and postcondition specification if it is partially correct (it is safe) and it terminates (it is always live, i.e. it is never the case that it does not produce any results).*

An important extension provided by the predicate transformers semantics is offered by the following schema:[34]

$$(wp(P, Q) \vee wp(P, R)) \rightarrow wp(P, Q \vee R)$$

i.e. the ability to account for non-deterministic machines, which in turn means that design starts usually from the specification in terms of post-conditions. In the case of a deterministic program P and some postcondition R, each initial state is an element in the following partition of cases:

1. executing P leads to a state satisfying R;
2. executing P leads to a state satisfying $\neg R$;
3. executing P fails to terminate.

---

[32] See also Manna and McCarthy (1969); Manna (1971).
[33] Due to (Dijkstra, 1975, p.548).
[34] See (Dijkstra, 1976, p.19).

But this does not happen for a non-deterministic program, which requires a weaker notion of precondition including the possibility of non-termination. Accordingly, the notion of weakest precondition is adapted to include non-deterministic machines:[35]

Definition 90 (Weakest Liberal Precondition) *The weakest liberal pre-condition, denoted wlp(P, R), is the condition that characterizes the set of all initial states such that the execution of program P will result in one of the following cases:*

- *a final state satisfying a given postcondition R,*
- *a final state satisfying ¬R,*
- *a non-terminating process,*
- *a final state not satisfying either R or ¬R,*
- *either a final state satisfying R or a non-terminating process,*
- *either a final state satisfying ¬R or a non-terminating process,*
- *either a final state satisfying R or ¬R, or a non-terminating process.*

Since the early work on program correctness, therefore, there is a significant stress on the requirement that the object of logical analysis consists of rigorously formulated expressions, reflecting the program specification as well as the program states. This strengthens the view that properties of programs can (and in fact should) be formulated and proven mathematically. This is the approach that many theoretically oriented computer scientists took from the 1960s onward in determining whether a program satisfies an agreed specification for all possible inputs by means of mathematical proofs. For Dijkstra, for example, program correctness is a property similar to mathematical ones and the structure of a correctness proof should guide program design, in this reinforcing the attitude of programmers to treat their artefacts as mathematical objects:

> given the problem, the programmer has to develop (and formulate!) the theory necessary to justify his algorithm. In the course of this work he will often be forced to invent his own formalism.[36]

The complication proper to programming is due to its inner physical nature, the lower level of abstraction it involves, which requires the programmer to deal not only with precision and explicitness but also with properties like complexity, e.g. in terms of memory size. This aspect was not left unattended even by these mathematically oriented supporters of program correctness, as it is clear from the following passage still due to Dijkstra:

> most techniques for proving the correctness of a program treat the program text as a static, rather formal, mathematical object that can be dealt with independently of the fact that there may exist machines that could execute such a program. As such, a clear separation of concerns emerges: we might call them the mathematical concerns

---

[35] Extracted from Dijkstra (1976).
[36] (Dijkstra, 1974, p.611).

about correctness and the engineering concerns about efficiency. In contrast to the correctness concerns, the efficiency concerns are only meaningful in relation to implementations and it is only during the efficiency concerns that we need remember that the program text is intended to evoke computational processes. Both the mathematical and the engineering concerns have, of course, always been with us, but once they used to be dealt with inextricably intertwined.[37]

The semantic approach

To establish a precise correspondence between programs and mathematical entities entirely independently of implementation was also the task of the denotational semantics developed by Scott and Strachey in the 1960s.[38] This method defines programs in terms of denotations of constructs expressing functions and representing information available before and after its execution. The focus is on the semantic evaluation of an abstract syntax defined for correct constructions by variables Var, expressions Exp, commands Cmd, and programs Prg in BNF form with composition, case construction, assignment and appropriate extensions by fixed-points of continuous functions for recursion and loops. Semantic functions mapping expressions to their denotations are constructed inductively from syntactic constructs, a set S of program states σ and positive integers $\mathbb{N}$ for values:

Definition 91 (Syntax of functions)

$$M : \text{Prg} \rightarrow (\mathbb{N} \rightarrow \mathbb{N})$$
$$E : \text{Exp} \rightarrow (S \rightarrow \mathbb{N})$$
$$\sigma : S = \text{Var} \rightarrow S$$
$$C : \text{Cmd} \rightarrow S \rightarrow S$$

Then for every $\Gamma \in$ Cmd, the expression $C[\Gamma]$ denotes the state transition specified by the command Γ and $(C[\Gamma])\sigma$ is the state after the execution of Γ, given σ is the state before its execution.[39] Denotational semantics explicitly evaluates formulas compositionally through correct syntactic constructions, but the focus is on the semantic level. Such precise definition in agreement with the standard constructs of real programs represents a semantic expression of criteria of correctness for implementations, without need for determining them. The operation of binding identifiers to appropriate denotation is explicitly traced back to Church's λ-calculus through the functional application operation, which is the primitive construct of any binding (syntactic sugaring). And among the other problems to be addressed in this model, along with the definition of recursion, there is the need to restrict self-application of higher-order functions which can generate paradoxical constructs entirely similar to those of set theory.

---

[37] (Dijkstra, 1975, p.549).

[38] See Scott and Strachey (1971); Tennent (1976); Milne and Strachey (1977) for the original works and Mosses (2006) for a complete overview of the field.

[39] This formulation is extracted with some simplification from (Tennent, 1976, pp.438–9).

The resource-based analysis

In the last two decades, the formal approach to proving program correctness has received a huge boost from the interpretation of the problem at a lower level of abstraction, namely resources management at execution. This approach addresses the need to prevent errors in the way software memory is used. This approach is at the basis of the most recent logical framework to address the problem of verification in computer science: separation logic.[40]

A reason for the success of separation logic is its ability to mirror the dual abstract-physical nature of software systems, an aspect we will insist on in the second part of this volume. Moreover, its proof-theory is apt to implementation in verification tools scalable to large software systems. The main issue resolved by this approach is the reliable allocation of resources, and in particular random access memory and task scheduling. Separation logic models have pointers and memory locations as primitives. Proof-theoretically this is done first in a logic of bunched implications (**BI**, see Figure 7.9), which expresses the principle of composable, finite resources. From the point of view of resources, a crucial step is the distinction between the additive and multiplicative forms of conjunction, originally due to linear logic.[41] The connectives become distinct if the rules of weakening and contraction are missing, making the logic substructural. Formally, this is expressed by the different notations ; and , used to construct contexts on the left-hand side of the derivability sign $\vdash$. At the same time, the system reflects the engineering nature of the computational artefacts it models by expressing reasoning about the resources through a connective $*$ which expresses multiplicative conjunction ('and, separately').

This formal apparatus is technically completed by the use of pre- and postconditions expressions inherited from Hoare Logic in the semantics. This happens through the modelling of resources R in two elements $\langle s, h \rangle$:

- the stack s refers to the ordered array of data on top of which elements are put and taken off by variables mapping to values, i.e. $s \in S$ with $S = Var \rightarrow_{fin} Val$ a set of finite partial functions from variables to values;

- the heap h refers to the ordered array of data accessible at any point as long as a pointer with an address is provided through mapping addresses into values, i.e. $h \in H$ with $H = Loc \rightarrow_{fin} Val \times Val$, i.e. a set of finite partial functions from locations to (pairs of) values.

The stack and the heap are physical entities formally represented by provable sentences with consequences. The main forcing relation $s, h \models \phi$ asserts that the atomic formula ϕ is true of stack and of the heap. In the semantics, the heaps are the worlds in which formulas are evaluated; see Figure 7.10. In this language, E stands for an expression and $[[E]]_s$ is an heap-independent value; the points-to relation $\mapsto$ is used to make statements

---

[40] For technical foundational papers on separation logic, see O'Hearn and Pym (1999); Ishtiaq and O'Hearn (2001); Reynolds (2002). For an accessible introduction to the way this logic stresses both the formal and physical model of computational artefacts, see Pym et al. (2019).

[41] See Girard (1987).

$$\frac{}{\phi \vdash \phi} \ \text{Identity} \qquad \frac{\Gamma \vdash \phi}{\Delta \vdash \phi} \ \Gamma \equiv \Delta, \equiv$$

$$\frac{\Gamma(\Delta) \vdash \phi}{\Gamma(\Delta; \Delta') \vdash \phi} \ \text{Weakening} \qquad \frac{\Gamma(\Delta; \Delta) \vdash \phi}{\Gamma(\Delta) \vdash \phi} \ \text{Contraction}$$

with $\Gamma(\Delta)$ a bunch Γ in which Δ appears as subtree.

$$\frac{}{\{\}_m \vdash I} \ \text{I-I} \qquad \frac{\Gamma(\{\}_m) \vdash \chi \qquad \Delta \vdash I}{\Gamma(\Delta) \vdash \chi} \ \text{I-E}$$

with $\{\}_m$ the multiplicative unit and I the empty resource (in the multiplicative fragment).

$$\frac{\Gamma, \phi \vdash \psi}{\Gamma \vdash \phi \mathbin{\text{--}\ast} \psi} \ \text{--}\ast\text{-I} \qquad \frac{\Gamma \vdash \phi \mathbin{\text{--}\ast} \psi \qquad \Delta \vdash \phi}{\Gamma, \Delta \vdash \psi} \ \text{--}\ast\text{-E}$$

$$\frac{\Gamma \vdash \phi \qquad \Delta \vdash \psi}{\Gamma, \Delta \vdash \phi \ast \psi} \ \ast\text{-I} \qquad \frac{\Gamma(\phi, \psi) \vdash \chi \qquad \Delta \vdash \phi \ast \psi}{\Gamma(\Delta) \vdash \chi} \ \ast\text{-E}$$

$$\frac{}{\{\}_a \vdash 1} \ \text{1-I} \qquad \frac{\Gamma(\{\}_a) \vdash \chi \qquad \Delta \vdash 1}{\Gamma(\Delta) \vdash \chi} \ \text{I-E}$$

with $\{\}_a$ the additive unit and 1 the empty resource (in the additive fragment).

$$\frac{\Gamma; \phi \vdash \psi}{\Gamma \vdash \phi \to \psi} \ \to\text{-I} \qquad \frac{\Gamma \vdash \phi \to \psi \qquad \Delta \vdash \phi}{\Gamma; \Delta \vdash \psi} \ \to\text{-E}$$

$$\frac{\Gamma \vdash \phi \qquad \Delta \vdash \psi}{\Gamma; \Delta \vdash \phi \wedge \psi} \ \wedge\text{-I} \qquad \frac{\Gamma(\phi; \psi) \vdash \chi \qquad \Delta \vdash \phi \wedge \psi}{\Gamma(\Delta) \vdash \chi} \ \wedge\text{-E}$$

$$\frac{\Gamma \vdash \phi_i}{\Gamma \vdash \phi_1 \vee \phi_2} \ (i = 1, 2), \vee\text{-I}$$

$$\frac{\Gamma \vdash \phi \vee \psi \qquad \Delta(\phi) \vdash \chi \qquad \Delta(\psi) \vdash \chi}{\Delta(\Gamma) \vdash \chi} \ \vee\text{-E}$$

Figure 7.9 Rules of BI Logic, O'Hearn and Pym (1999)

about the contents of heap cells; $dom(h)$ denotes the domain of definition of a heap h and $dom(s)$ is the domain of a stack s; $h\#h'$ indicates that the domains of heaps h and h' are disjoint and $h \cdot h'$ denotes the union of disjoint heaps.

These new models are used to predict behaviour rather than to establish which artefacts compute and which not. There is a feedback loop between what the logic can prove concerning possible errors on the resources used by a section of a program and what the

$s, h \models E = E'$ iff $[[E]]_s = [[E']]_s$

$s, h \models E \mapsto E_1, E_2$ iff $\{[[E]]_s\} = dom(h)$ and $h([[E]]_s) = \langle [[E_1]]_s, [[E_2]]_s \rangle$

$s, h \models emp$ iff $h = []$

$s, h \models P * Q$ iff $\exists h_0, h_1. h_0 \# h_1, h_0 \cdot 1 = h, s, h_0 \models P$ and $s, h_1 \models Q$

$s, h \models P \twoheadrightarrow Q$ iff $\forall h'$. if $h' \# h$ and $s, h' \models P$ then $s, h \cdot h' \models Q$

$s, h \models false$ never

$s, h \models P \Rightarrow Q$ iff if $s, h \models P$ then $s, h \models Q$

$s, h \models \exists x.P$ iff $\exists v \in Val.[s \mid x \mapsto v], h \models P$

Figure 7.10 Semantic evaluation of expressions in separation logic, Reynolds (2002)

empirical testing of the program can show about those proofs.[42] Errors in memory management, which such logical models aim at anticipating, concern the use of a null pointer where a valid value is required, losing track of memory items, security vulnerabilities like resource, and memory leaks.[43]

The approach underlying a formal system like separation logic allows one to reason directly about physical systems, designing an overlap between the abstraction and the implementation layers of computational systems. Thus, while the principled position of supporters of formal correctness à la Hoare is still subsumed in the slogan *Well-specified programs don't go wrong*,[44] here the specification has a much stronger requirement, in that it has to quantify both over the heap, the stack, and all possible executions, making its connection to implementation essential. In this context, conditions for program execution and termination are expressed in terms of resources $(R = \langle s, h \rangle)$ sufficient for the truth of the proposition expressing the intended output; this requires in turn the distinction between additive and muliplicative resources, which amounts to establishing which context admits structural rules of weakening and contraction, i.e. expressing contexts of resources which can be shared or that must be separated. The restriction of failing execution is possible thanks to the formulation of a semantics with elements $\langle C, R \rangle$ with C for code, R for resources, in terms of the logic of bunched implications, which has brought to a fault-avoiding interpretation of safety:

Definition 92 (Safety Yang and O'Hearn (2002)) $\langle C, R \rangle$ *is safe if there is no transformation according to C of the memory configuration given by R into a faulty memory configuration R'.*

This, in turn, restores the notion of partial correctness for Hoare triples (where ϕ and ψ express respectively pre- and postconditions):

---

[42] The use of the scientific method of model-building is in this context similar to what happens in the experimental sciences, a topic that will be explored at length in the third part of this volume.

[43] See (Pym et al., 2019, sec.2).

[44] Ishtiaq and O'Hearn (2001), paraphrasing Milner.

Definition 93 (Partial Correctness) $\{\phi\}C\{\psi\}$ *is true if for all* $R = \langle s, h\rangle$ *it is the case that* $R \models \phi$ *implies that* $\langle C, R\rangle$ *is safe and if there is a transformation according to C to a memory configuration* $R' = \langle s', h'\rangle$, *then* $R' \models \psi$.

Definition 94 (Total Correctness) $\{\phi\}C\{\psi\}$ *is true if for all* $R = \langle s, h\rangle$ *it is the case that* $R \models \phi$ *implies that* $\langle C, R\rangle$ *terminates normally and if there is a transformation according to C to a memory configuration* $R' = \langle s', h'\rangle$, *then* $R' \models \psi$.

A standard way to exemplify the different approach between Floyd-Hoare and separation logic is by stressing how the safety requirement excludes some trivial validity, e.g. an expression of the form `{true}C{true}` does not hold for every command.[45] The validity relation of interest holds between states of the program satisfied under changing resources configurations, while it notably requires adding either failure exclusion (for partial correctness) or termination (for total correctness). The soundness and completeness results for this logic establish the coincidence of the logical and engineering models.[46]

The debate that started towards the end of the 1970s on the formal correctness research program would concentrate precisely on the different levels of abstraction that programs as abstract and implemented objects require for their definition and the identification of their properties.

7.3 The Debate

Across the 1960s, the main objective of the newly born discipline of computer science is seen by many in logical terms as symbol manipulation, focusing on how properties essential to computing (e.g. finiteness) would impact on their logical and mathematical representation. The approach underlying the proofs-as-programs correspondence, based on a strong mathematical and logical understanding of languages, had the advantage of guaranteeing that only *correct programs* can be generated by proofs: a theorem is extensionally equivalent to the intended problem specification.[47] The specification is here assumed and the proof construction method corresponds to constructing an element in the set which is expressed by such specification. There are at least two important issues and limitations of this understanding which are worth highlighting:

1. there are properties of programs that (at least at the very beginning of this tradition) could not be easily translated by constructions: partiality, concurrency, time control, AI methods;

2. the formulation of an algorithmic translation of a valid proposition requires a property of *efficiency*, as by Definition 64.

---

[45] See (Reynolds, 2002, sec3.3).

[46] See (Pym et al., 2019, sec.4.3). This is an aspect of crucial importance for the notion of validity in physical computing systems as explored in the second part of this volume.

[47] See e.g. (Constable, 1983, p.107).

The discussion that would ensue in the following decades concerning the design of provably correct programs would alternate between these two essential aspects of programs or, as we have called them above, these two levels of abstraction: programs can be represented as mathematical structures enjoying formal properties, but at the same time they require to be realized as physical artefacts. As we have illustrated above, this dichotomy would eventually guide the technical work in program verification. But before that, it also stirred a much larger debate on the philosophical aspects of computer science.

One of the earliest and most critical contributions which kickstarted this debate was due to De Millo, Lipton, and Perlis.[48] In an article directly addressed to the computer practitioners and professionals, the authors strongly rejected the thesis that a proof of correctness fully establishes confidence in the reliability of a program. Their argument relies on stretching the comparison between proofs and programs to the practice within the respective communities. In the case of proofs, they maintain, is the social acceptance of correctness by the community of mathematicians that establishes such confidence, rather than the proof in itself; this is in particular due to the fact that a proof can be opaque and incorrect in the first place. A proof, which includes both formal and informal aspects to it, is only partially equivalent to deductive reasoning, and often it can only be taken as a probable attempt at establishing the truth of a theorem. This, in itself, is obviously a claim that would be strongly rejected by authors with a strong mathematical and logical background. Instead, de Millo, Lipton, and Perlis argue, it is the presentation of the proof to the community that consolidates its veracity, in terms of its attractiveness and its ability to convince. Confidence in the truth of mathematical statements is obtained through proofs only when these are subject to the social mechanisms of the mathematical community. On the other hand, formal verification of program correctness has no such social process in place and it is therefore bound to fail. This is further justified by several elements:

1. verifications cannot be surveyed, as they are impossible to be shared among specialists;

2. the process of transforming an informal specification of the program to its formal counterpart for verification cannot be itself but informal, and as such it undermines the very purpose of the verificationist position;

3. formally verifying that a program is consistent with its specification requires that the two elements (program and verification) be independently derived: but the real design process of programs is often a feedback cycle between specification elicitation and implementation.

The first argument was probably the hardest to accept for formally-inspired computer scientists: the entire effort of verification, going back to the work of Turing, Goldstine and von Neumann, Floyd, was precisely that of offering not only an interpretation of programs as mathematically provable list of statements but in fact as surveyable ones,

---

[48] See De Millo et al. (1979).

especially through the design and use of graphical aids like flowcharts. Nonetheless, this problem was going to be crucial and of increasing relevance in all the later incarnations of verificationism, in particular with the exponential growth of software systems and the use of automated means to prove correctness.

The second critique was certainly more likely to be accepted by the formalists, who were very aware of the problem of moving from the abstract formulation of the program to its implementation, and then to the execution level.[49]

The latter claim was undoubtedly the hardest to counter, and the problem of the definition loop between specification and implementation was one that would return not only as a major aspect of software engineering but also in the philosophical debate of the later decades.[50]

On this basis, the authors dismiss the idea of realizability of full formal verification of complex, real software systems, as these do not scale continuously from smaller programs, for which on the contrary verification is actually possible. This all reduces to the belief that working with software systems means working with systems one does not understand, and not create perfectly: the only principle that can be invoked, in place of verifiability, is reliability, i.e. systems which can be believed to work *sufficiently well*.[51]

The critique to formal verification was thus that proving correctness would not imply necessarily specification adherence. Another critical voice in this direction was Brian Cantwell Smith:

> there are inherent limitations to what can be proven about computers and computer programs. [...] Just because a program is 'proven correct', in other words, you cannot be sure that it will do what you intend.[52]

The main argument by Cantwell Smith is devoted to the use of models in program construction: these models are defined at some level of abstraction and as such they are partial. On the contrary, the result of actions (including the results of running a program: landing aircrafts, administering drugs, and similar) are not partial. The action (and the computer running a program) can therefore be used to provide feedback to complete the model that we know is not correct (because it is partial). Hence a program is as correct as the model is, and this is always some step away from the correctness of the real world. The program developed will differ from the specification in that it has to say how the intended behaviour is to be achieved (i.e. provide the system's functional requirements), while the specification only requires to declare what the proper behaviour will be (i.e. the non-functional requirements). Then a proof that a program satisfies a given specification is just a way to show that two characterizations of a given behaviour (functional, non-functional)

---

[49] This is clearly witnessed by the quote from Dijkstra (1975) presented above pp. 95–96.

[50] This will be explored at length in Chapter 12.

[51] It is worth anticipating here that the property of minimal well-functioning will be crucial in our assessment of both physical computing in Part II and of experimental computing in Part III of this volume.

[52] (Cantwell Smith, 1996, p.276).

are compatible; but such compatibility does not guarantee that the specification is in itself a correct representation of your intention:[53]

Definition 95 (Relative Consistency) *A proof of the compatibility or consistency between two formal objects of an extremely similar sort: program and specification.*

The value of verification lies therefore in making your program design a lot clearer and to discover bugs that would otherwise go undetected.

Another step towards the weakening of the appropriate notion of correctness is due to an highly debated contribution by Fetzer.[54] While criticizing the view by De Millo, Lipton, and Perlis, Fetzer radically opposes the idea that programs can be treated as logical structures subject to deductive verification. Relying on the principle of analyticity of deductive consequence (the truth of the conclusion is contained in the truth of the premises if the inference is to be correct), Fetzer qualifies correctness of a program P in two forms. On the one hand, there exists some program P interpreting certain rules of inference or program expressing primitive axioms which permit one to draw inferences concerning the performance of a machine executing P.

Definition 96 (Absolute Verifiability or Validity) *Within a system of natural deduction or an axiomatic formal system, the members of the class of consequences that can be derived from no premises at all or that follow from primitive axioms alone may be said to be absolutely verifiable.*

On the other hand, there exists a set of premises concerning P from which it can be derived a conclusion concerning the performance of a machine executing P.

Definition 97 (Relative Verifiability or Derivability) *Those members of the class of consequences that can only be derived relative to specific sets of premises whose truth is not absolutely verifiable may be said to be relatively verifiable.*

In developing this dual conception of verifiability, Fetzer suggests that a model of proof that can account for errors needs to be inherently probabilistic. Verification, according to Fetzer, can then work for small programs (in the vein De Millo et al. (1979) argued for) and for large programs when they consist of relatively straightforward arrangements of smaller programs (exhibiting what he calls *cumulative complexity*). On the other hand, proving correctness is more difficult for larger programs consisting of complicated arrangements of smaller programs (exhibiting what he calls *patch-work complexity*).

A program, after all, is a particular implementation of an algorithm in a form that is suitable for execution by a machine. In this sense, a program, unlike an algorithm, qualifies as a causal model of a logical structure of which a specific algorithm may be a specific instance. The consequences of this realization are enormous, insofar as causal models of logical structures need not have the same properties that characterize those logical structures themselves. Algorithms, rather than programs, thus appear to

---

[53] This is the same issue already highlighted by De Millo et al. (1979).
[54] See Fetzer (1988).

be the appropriate candidates for analogies with pure mathematics, while programs bear comparison with applied mathematics. Propositions in applied mathematics, unlike those in pure mathematics, run the risk of observational and experimental disconfirmation.[55]

Hence, program verification is an inappropriate and unjustifiable translation of a deductive procedure applicable to theorems and to algorithms for the purpose of evaluating causal models that are executed by machines. Fetzer is not entirely dismissive of the aims of formal verification, but this task needs to be formulated at the right level of abstraction, clearly distinguishing algorithms from programs (in the same way it was suggested in Chapter 6). The term *program* may be used to refer to:

- algorithms;
- encodings of algorithms;
- encodings of algorithms that can be compiled;
- encodings of algorithms that can be compiled and executed by a machine.

Programs as algorithms and their encodings are absolutely verifiable. Programs as compilation and execution are only relatively verifiable with respect to given inputs and outputs (which are not deductively, but only inductively provable). The equivocation of formal verification is therefore based on the idea that its theoretical possibility holds for both algorithms and their encodings (whose behaviour can be proven with certainty), as well as compiled and executed code (whose behaviour can be proven only probabilistically). Programming as a mathematical activity holds only for formal expressions, while it does no longer hold for programs understood as objects of an applied mathematical activity.

It is worth noting how the separation of levels of abstraction was abandoned in later years by supporters of the strong approach to formal verification. While some of the proponents of the formalist view like Dijkstra seemed to agree upon the need to distinguish between algorithms or textual programs and their impemented versions, in a talk given in 1986, Hoare maintained four basic tenets of his view on verification:[56]

1. Computers are mathematical machines;
2. Computer programs are mathematical expressions;
3. A programming language is a mathematical theory;
4. Programming is a mathematical activity.

This view implies that the behaviour of a machine running a compiled program is precisely derivable by knowing the initial states of the machine and the precise description of the program. Note that this is a much stronger claim than the one assuming the formal description of the program to be fully deterministic, and it certainly suggests a

---

[55] (Fetzer, 1988, p.1057).
[56] See (Hoare, 1993, p.135).

collapse of the levels of abstraction between algorithm, textual program, and compiled program. This list of principles, nonetheless, is followed by important provisos: first, the physical architecture of machines is complex and often poorly defined; second, computers have size of which mathematics has no previous experience; programming languages are complex; programmers are not mathematically trained. In this context, Hoare realizes that

> the engineer uses experience and understanding to formulate and apply various rules of thumb, and uses flair and judgment to supplement (though never to replace) the relevant mathematical calculations. Experienced programmers have developed a similar intuitive understanding of the behavior of computer programs, many of which are now remarkably sophisticated and reliable.[57]

The introduction of mathematical methods is suggested here both for its scientific value and for its economical return, to improve the status of programming as a discipline. This proposal seems to have been at least partly realized in modern verification.

After this initial, quite strong debate within the discipline itself, the principal actors seem to have gone quiet, and the field silently divided among those still believing in the role of formal methods and their development to verify programs correct, and the sceptics.[58] Today, formal verification is a much more mature field, which is gaining strong results, as illustrated by the use of formal tools like separation logic in large web-based software systems. Towards the end of the 1990s, results from formal methods of the previous 20 years looked obscure, unscalable, inadaquate, hard to use, trivial. On the other hand, the immediate future looked brighter, especially in the interaction of academia and industry through the following specialized disciplines:[59]

- formalization of specifications;[60]
- development of model-checking techniques in hardware and protocol verification, consisting in building a finite model of a system and checking that a desired property holds through exhaustive state space search guaranteed to terminate;[61]
- automated and interactive theorem proving for the mechanical verification of safety-critical properties of hardware and software designs, in which the system and its

---

[57] (Hoare, 1993, p.150).

[58] For how the debate was illustrated within the computing community, see Ashenhurst (1989).

[59] This analysis is presented in Clarke and Wing (1996), an early survey on the literature and industrial development in formal methods.

[60] For example with the development of the Z and VDM languages, see respectively Spivey (1988) and Jones (1986).

[61] This was in particular exemplified by temporal model checking and finite state transition systems; see Baier and Katoen (2008).

desired properties are expressed as formulas in some mathematical logic and the process consists in finding a proof of a property from the axioms of the system.[62]

At the dawn of the second decade of the twenty-first century, the range of tools available for formal verification are extensive: static analysers identifying potential program defects, model checkers and abstract interpreters, and theorem provers, each category presenting a specific trade-off between the level of automation offered and the complexity of the analysis performed.[63] A list of projects in which verification is used includes microprocessors for parallel processing, train signals control, smart-card-based cash system, airflight controllers, control of a movable barrier for flood protection, biometrics access control, firmware for integrated circuits cards; future pilot projects include: verified file store, verified mini-kernel for an OS, radio spectrum auctions management, cardiac peacemaker, verification of the Microsoft Hypervisor (for multiple guest OSs on a single hardware).

A revised, stronger philosophical debate has also emerged from the modern realizations of verification methods and their properties, in particular by the development of automated theorem proving. In these later analyses of the problem, arguments can be identified against or in favour of the possibility of proving correctness.[64] One category of arguments can be labelled *epistemological*:

Thesis 6 (Epistemological Argument 1) *The result of a computation by a machine is justifiable empirically (hence only a posteriori), because evidence for the existence of a formal proof mapped by the steps of a program executed by a machine presupposes the reliability of that machine, and due to complexity reasons the knowledge that computers are reliable can only be justified empirically.*[65]

Thesis 7 (Epistemological Argument 2) *The result of a computation by a machine is justifiable a priori, i.e. it is warranted neither on sense experiences nor on sense-perceptual beliefs, because an individual's knowledge of pure mathematics is a priori; because computers are not autonomous thinkers and their manipulation of mathematical symbols is meaningful only insofar they are programmed; and because mathematical knowledge is a priori and it can be transmitted through communication.*[66]

Thesis 8 (Epistemological Argument 3) *Any logical argument that aims at refuting verifiability in the style of Fetzer's argument in order to establish that program verification is both theoretically and practically impossible is self-defeating: the argument in question can be checked mechanically and, in turn, proof checking is a valid interpretation of program*

---

[62] This was exemplified by processor design checkers (Kuehlmann et al. (1995)), binary machine code checkers (Boyer and Yu (1996)), and proof checkers in particular for program verification (Gordon (1987); Luo and Pollack (1992); Team (2017)).

[63] Woodcock et al. (2009) highlighted the practice at 20 years distance from the previous overview, illustrating the attempt at gaining further understanding in the various trends of the discipline.

[64] For an overview of some of these arguments, see Turner and Angius (2017).

[65] See Tymoczko (1979, 1980).

[66] See Burge (1998).

verification (note that here one is considering the decidable task of checking a proof, rather than the undecidable one of finding a proof); then if the argument refuting verifiability holds, it can be checked and therefore verification must be both practically and theoretically possible; by contradiction the argument is self-refuting.[67]

The first argument is essentially against the idea of the possibility of proving correctness in principle; the second and third arguments are essentially in favour. Another set of arguments can be labelled *ontological*, stressing the essential role that physical artefacts have in determining what computing actually is; in this category we identify two main arguments:

Thesis 9 (Ontological Argument 1) *A physical artefact S can correctly implement its abstract specification C if there is a mapping from the states of S to the states of C such that the transitions between states in S mirror the state transitions in C (simple mapping account); as the mapping is extensional, if unrestricted it makes any system an implementation of another by justaxposition of the corresponding elements;*[68] *this makes correctness with respect to a specification hard to establish, as syntactic correctness is no longer sufficient, while the semantics of its function needs to be determined in the larger context of the system hosting the program.*[69]

Thesis 10 (Ontological Argument 2) *Determining correctness of a physical artefact S against an abstract specification C assumes that S is well-functioning; to establish this to be the case, one can only check for the absence of errors and a taxonomy has to be formulated at various levels of abstraction: the distinction of different types of miscomputations determine the limits of what formal verification can account for.*[70]

While the ontological arguments do not need to be interpreted as strictly in favour or against the possibility of formal correctness in general, it is obvious that they stress the value of levels of abstraction in the analysis of computational artefacts, thereby also aiming at determining the proper roles of specification and implementation and the different levels of correctness for each.

Let us now propose some final observations on the notions of validity and correctness for the mathematical foundation of computing.

7.4 Formal Computational Validity

The mathematical foundation of computing, its interpretation as a science of algorithm, and the formal identity of programs and proofs have been a constant component in

---

[67] See Arkoudas and Bringsjord (2007); Bringsjord (2015).

[68] The simple mapping account is explored at length in Piccinini (2015); we will return to its formulation in Chapter 10.

[69] For this argument, see also Turner (2012).

[70] Such a taxonomy is presented in Fresco and Primiero (2013) and will be reviewed in Chapter 10.

Level of Abstraction	Computing Notion	Information
Abstract Level	Algorithm	Abstract Information
Implementation Level	Program	Instructional Information
Execution Level	Compiled Program	Operational Information

Figure 7.11 Abstraction levels in computing

the attempts made at offering an appropriate definition of computing and of computer science as the academic discipline most closely related to it. Such an approach, though, has progressively been balanced by the counterpart interpretation stressing the physical nature of computing as machinery, which will be investigated at length in the next part of this volume. We take this emerging dichotomy as both a conceptual and historical starting point to investigate the debate on the relation between computer science and mathematics.[71] We aim at illustrating the progressive emerging of an understanding of computing as a mathematical discipline. One way to approach this analysis is methodological, through the identification of the appropriate levels of abstraction at which computing artefacts can be understood mathematically.

The methodology of the levels of abstraction refers to a standard approach in computer science (and software engineering in particular), which usually points to a technique that takes only certain aspects of a system under consideration, hiding information that is not considered relevant in a particular setting. Levels can, for example, refer to the physical structure, the logical design, or the entire system respectively. This technique has been adopted in epistemological analyses at large through the philosophy of information.[72] It formally consists in adopting a set of typed variables equipped with an equality relation, an interpretation on such variables called *observables* and it defines a level of abstraction as a finite, non-empty set of such interpreted terms.[73] In the present analysis—and in the rest of this volume—we will refer to the levels of abstraction in computer science as determined by the interpreted variables that are taken as objects of study by the discipline: in other words, we are interested in the formalization level of the notion of computing. This can be also expressed in terms of the corresponding notion of information at stake at each level of abstraction.[74] This representation of the levels of abstraction in computer science has already emerged in our analysis of algorithms, and shall remain of reference in the remaining part of this volume; see Figure 7.11.

In particular, we require an association of types of information with each level of abstraction and notion of computing. At the highest LoA, computing is understood as

---

[71] This analysis will be extended in this volume with the analysis of the relation between computer science and engineering on the one hand, and computer science and natural sciences on the other, respectively at the end of Parts II and III.

[72] See (Floridi, 2011, ch.3).

[73] See in particular (Floridi, 2011, p.52) and also Floridi (2014).

[74] See Primiero (2016).

a mathematical science, whose objects of study are algorithms and where the content of the latter is abstract information:

Definition 98 (Abstract Information) *The content of an algorithm is abstract, correctness- and truth-determining information.*

The abstract nature of this information is expressed by the fact that it is linguistically dependent, but encoding independent up to equivalence of the computability class. Moreover, algorithms determine which implementations are correct, and which models are appropriate, hence which are true instances. This is the content of the abstract objects considered in Chapter 6. It is abstract because it should be independent of any language-specific syntax; and it is semantic because it has to express the intended meaning of the system, i.e. it will describe what the system is supposed to do.

We introduce here for completeness the definitions of information at lower levels of abstraction. Computing intended as a science of programs requires an instructional notion of information:

Definition 99 (Instructional Information) *The content of a program is instructional, well-formed, meaningful data.*

Syntactic well-formedness is defined over abstract objects and their properties, such as integer numbers and the sum operation. Meaning is acquired at this stage by evaluating whether an operation (and in turn the action it determines) is obtained at the implementation level (i.e. operationally, not denotationally). There is still no alethic assessment: like in the case of an order to a person, it makes no sense to ask of a piece of code in itself whether it is true or false.

Finally, computing intended as a science of machines subsumes an operational notion of information:

Definition 100 (Operational Information) *The content of machine code is structured, physical, performative data as evaluated variables.*

In this case we refer to performance of actions on the physical hardware, where well-formedness is essential for execution and information emerges from the physical and signal difference that variable evaluation produces.[75]

In one of the earliest definitions of computer science from 1967, Newell, Perlis, and Simon approach this tentative balance between abstraction, implementation and execution:

> computer science is the study of computers. The phenomena surrounding computers are varied, complex, rich.[76]

---

[75] The ontological and epistemological aspects associated with each these notions and their combination are explored in Part II of this volume.

[76] (Newell et al., 1967, p.1373).

This attempt seems to stress the physical level of machines, executable programs and operational information, but while it starts from our bottom layer it also aims at preserving the multi-faceted nature of the discipline as an hybrid between a mathematical and an engineering discipline. The authors analyse various objections to their focus on machines: of these, the strongest is probably that computer science is the study of algorithms (or programs), not computers, or in other words that computing sits at a higher level of abstraction. The authors respond with an understanding of computers as machinery that is designed and works according to mathematical rules, i.e. *computers plus algorithms*. Note that the formulation of this objection suggests, at this point in time, a merging of the notions of algorithm and program which we have kept distinct as, respectively, abstract and instructional information.

In the same year, Forsythe defines computer science as follows:

> the art and science of representing and processing information and, in particular, processing information with the logical engines called automatic digital computers.[77]

Here a separation of levels of abstraction is made clear: the design and representation of computational problems are based on algorithms and symbol manipulation; the processing of information on machines focuses on optimization aspects, a central engineering aspect. In a new attempt made a year later, Forsythe stresses that computer science shares with mathematics the goal of creating basic structures whose concepts are independent of applications. Nonetheless, the algorithmic aspect of the discipline is strongly linked to a pragmatic aspect, defined by the dynamic of computation. He further went into discussing in how far computer science can be understood as applied mathematics, while sharing at the same time the application aspects with engineering. Computer science and mathematics share, according to Forsythe, the need for

> rigor and exposition (in mathematicians' language), or performance and documentation (in computer science terminology), and place a higher premium on quality than on promptness.[78]

Finally, he likens digital computers to experimental tools for applied mathematics.[79]

These two early attempts at defining the discipline of computer science already illustrate the dual nature that practitioners aware of the mathematical foundations were trying to preserve. During the 1970s, this duality of computer science between mathematics and engineering was still unresolved. Knuth describes computer science as

---

[77] (Forsythe, 1967, p.3).

[78] (Forsythe, 1968, p.457).

[79] This specific aspect is one to which we will return in later interpretations of the discipline and which we will explicitly address in the Part III of this volume.

> the study of algorithms [...] a precisely defined sequence of rules telling how to produce specified output information from given input information in a finite number of steps. A particular representation of an algorithm is called a program.[80]

Here it is important to stress how Knuth conceptually separates the algorithm from its representation, followed by computers as the mechanical implementation of programs. At this stage, therefore, the basic distinction between levels of abstraction mentioned above is fully determined. While algorithms are singled out as the core of the discipline, common to its many branches, it is interesting to note how Knuth formulates the need for the mechanical execution of programs:

> But computers are really necessary before we can learn much about the general properties of algorithms; human beings are not precise enough nor fast enough to carry out any but the simplest procedures. Therefore the potential richness of algorithmic studies was not fully realized until general-purpose computing machines became available.[81]

In turn, computer science and mathematics have a very close and reciprocally influential relationship. Knuth however rejects the equivalence of the two disciplines, as essentially different viewpoints belong to them, with computer science still deemed too young to aim at the depth of results available in mathematics. Moreover, they present an essential difference in their methods: while mathematics allows for infinite processes and static relationships to prove theorems, computer science allows for finitary constructions and dynamic relationships to formulate algorithms. Nonetheless, computer science is presented as a discipline which in the future would be considered basic to general education and no differently to what happened with mathematics,

> computer science will be somewhat different from the other sciences, in that it deals with man-made laws which can be proved, instead of natural laws which are never known with certainty.[82]

In the same journal issue, Dijkstra approaches the problem of the mathematical nature of programming.[83] Dijkstra starts by characterizing the different properties of mathematical work and programming; see Figure 7.12. A clear distinction between mathematics and computer science emerges from considering the latter to account for the practice of programming, and thus to include at least the implementation level. Nonetheless, in view of the need to rely on obedience when writing programs, the programmer is required to be accurate and precise, thus matching the first of the characteristics of work of a 'mathematical nature'. In view of the very nature of the algorithms that written programs

---

[80] (Knuth, 1974, p.323).
[81] (Knuth, 1974, p.323).
[82] (Knuth, 1974, p.326).
[83] See Dijkstra (1974).

Mathematical Work	Programming
Precision	Performance
Generality	Capacity
Rigour	Reliability on Obedience

Figure 7.12 Mathematics vs. programming, Dijkstra (1974)

implement, the programmer's work is also always general, as a program can be supplied with different inputs and always supposed to produce an appropriate result: hence, it matches as well the second of the listed characteristics of mathematical work. Finally, a program should be composed by well-designed components and if so, the resulting predictability guarantees the same confidence induced by mathematical work. Therefore, accounting for computing as both a science of programs and of algorithms allows one to establish the direct relation of computer science with mathematics. Note, however, how Dijkstra does not shy away from commenting on the weakest aspect of programs, namely their inherent unreliability: while at first this could have been ascribed to the fact that programming was still a young discipline, it then turned out to be an essential issue, only made worse by increasingly complex systems. The lowest of the levels of abstractions, that of execution of compiled programs, is thus the one which appears to test the mathematical foundation of computing the most.

With respect to the highest level of abstraction, though, programs inherit properties of proofs as their mathematical counterparts. In our analysis, programs intended as the counterpart of proofs correspond to algorithms as abstract machines feasible for implementation according to Definition 72 and they satisfy a principle of linguistic description:

Principle 5 (Linguistic Dependence) *Computation expressed by abstract machines is an algorithmic process in which a description of an input-output relation is linguistically encoded.*

Note that which encoding is chosen for the algorithm is irrelevant, as far as the level of abstraction is left unchanged:

Principle 6 (Encoding Independence) *Assuming the constraints of the chosen abstract machine, the output of a given equivalence class of computations is unaffected by the properties of the linguistic encoding.*

There is therefore a correspondence between

- computation by an abstract machine $(\mathcal{S}(A), \mathcal{I}(A), \tau_A)$ respecting linguistic encoding and implementability,
- its denotational semantic counterpart $(C[\Gamma])\sigma$,
- and computation by a syntactic term or program P within a context of resources $\mathcal{N}$ and its specification S.

At this level of analysis, syntactic correctness corresponds to checking whether a pair program-specification satisfies a certain logical relation of typing. This can be expressed by variations of well-known logical problems:[84]

Definition 101 (Type-Checking Problem) *Given a set of resources $\mathcal{N}$, a program P and a specification S, is it possible to derive a formula $\mathcal{N} \vdash P : S$ saying that the program P produces an output of type S when executed in the context of $\mathcal{N}$?*

The type-checking problem is effectively the program version of the decision problem, and by analogy with it, we know it to be semi-decidable, i.e. for every terminating program this assertion is actually verifiable. A weaker version is given by the problem of determining whether a program does exist that satisfies a certain behaviour:

Definition 102 (Type-Reconstruction Problem) *Given a set of resources $\mathcal{N}$ and a specification S, is there a program P such that it is possible to derive a formula $\mathcal{N} \vdash P : S$ saying that the program P produces an output of type S when executed in the context of $\mathcal{N}$?*

For the decidable case, the type-reconstruction problem is reducible to the type-checking problem, i.e. one can always execute some P in $\mathcal{N}$ to see whether S is satisfied by P and do that as long as it is needed to reach the appropriate output. It is instead undecidable for higher-order languages, like for typed languages with dependencies. This means that decidability is affected by the ability of establishing termination for all the processes in $\mathcal{N}$ required by P. Note, moreover, that in the type-checking problem above we presuppose knowledge of the type S, i.e. we assume that there is an intended output for the program P which can be verified against. This becomes a crucial aspect when considering the physical implementation of P (the main problem in the debate on formal correctness from the previous section), and the limits this has on well-functioning and in turn on correctness. Because it means that one needs to guarantee admissible conditions for P to be executed and to terminate, also depending on P being able to access the resources in $\mathcal{N}$, where each of these resources in turn needs to satisfy correctness conditions.

At the semantic level, the problem of formal correctness of computation by abstract machines can be expressed as one of compatibility between the models of the specification and the model of the program. In this sense, a semantic notion of computational validity expresses the realizability of the model of the program among those possible according to the specification.[85] We formulate therefore a notion of validity for computations in the following terms:

---

[84] For this analogy and some of the observations made in the following, see Primiero (2015).

[85] The problem of formulating the relation $\{R_1, \ldots, R_n\} \models S$ between requirements $\{R_1, \ldots, R_n\}$ and a system S for the latter to satisfy sufficient conditions for the former has been investigated informally in Floridi (2017). This approach has received a formal translation in Primiero (2019a), making explicit the system in terms of program P and its required resources and investigating efficient and optimal realizations. Note that $\{R_1, \ldots, R_n\}$ and S in the notation from Floridi (2017) are equivalent respectively to the specification S and the pair $(\mathcal{N}, P)$ in our interpretation.

Definition 103 (Valid Formal Computation) *Given a specification S, we say that a computation by a program P with resources $\mathcal{N}$ formally validates S in a given linguistic encoding L, denoted*

$$(\mathcal{N},P) \vDash_L S$$

if and only if there is no model $\mathcal{M}$ such that every expression e required by the linguistic encoding L of $(\mathcal{N},P)$ is true but S is false in $\mathcal{M}$.

The distinction between program and resources is instrumental in making it possible to formulate this definition more precisely in view of error-handling, requiring an understanding of malfunctioning for computational artefacts. These issues are at the core of the next part of our investigation, namely the engineering foundation.

Exercises

Exercise 67 *Define the Curry-Howard-de Bruijn isomorphism and explain its meaning.*

Exercise 68 *Explain the Browuer-Heyting-Kolmogorov semantics and how it contributes to the meaning of programs.*

Exercise 69 *What is, in abstract terms, the program correctness problem?*

Exercise 70 *Explain the semantics of programs by Floyd.*

Exercise 71 *Explain the notion of correctness by Hoare.*

Exercise 72 *Define partial correctness. Define total correctness.*

Exercise 73 *Explain the relation between program correctness and relative consistency.*

Exercise 74 *In which sense is program correctness unfeasible to be treated by deductive methods according to Fetzer?*

Exercise 75 *What are the inner differences in the social construction of mathematical truth and program correctness according to De Millo et al.?*

Exercise 76 *For which reasons is program correctness impossible according to the model analysis of Cantwell Smith?*

Exercise 77 *List the epistemological arguments against or in favour of formal verification.*

Exercise 78 *List the ontological arguments analysing the problem of formal verification.*

Exercise 79 *Define the main levels of abstractions in the definition of computation.*

Exercise 80 *Define syntactic correctness and semantic validity for computations.*

Part II
The Engineering Foundation

8 The First Generation of Computers

Summary

In this chapter, we start with the analysis of the engineering foundation of computing which, proceeding in parallel with the mathematical foundation, led to the design and creation of physical computing machines. We illustrate the historical evolution of the first generation of computing machines and their technical foundation, known as the von Neumann architecture. From the conceptual point of view, we clarify the relation between the universal model of computation and the construction of an all-purpose machine.

8.1 Shannon's Circuits

The engineering progress leading to the successful creation of physical computing machines goes along almost in parallel with the mathematical foundation. There are plenty of histories covering the development of various machines, technologies, and their inventors and it is not the aim of this and the following chapters to propose a new one.[1] Instead, we will consider the evolution of the technologies that have made those machines possible. We hope to provide a basic technical knowledge to appreciate the historical progress of the engineering foundation and to prepare the ground to understand some of the deep philosophical issues related to it.

To stress the parallel evolution of the mathematical and engineering foundations, it is worth starting by noting that one of the earliest crucial results for the realization of

---

[1] For a historical overview of computing covering most of the machines and their technologies, see e.g. O'Regan (2012). For a more detailed analysis of the architecture of first-generation computers, see Rojas and Hashagen (2002). For a complete overview of the technologies and their impact in a historical setting, see Ceruzzi (2003).

On the Foundations of Computing. Giuseppe Primiero, Oxford University Press (2020). © Giuseppe Primiero.
DOI: 10.1093/oso/9780198835646.001.0001

physical computers dates back to just one year after that *annus mirabilis* of the Church and Turing papers. In 1937, Claude Shannon in his Master's Thesis at MIT illustrates how to implement Boolean logic in the design of electrical circuits.[2] The problem consists in representing any circuit in terms of a set of equations corresponding to its relays and switches.[3] An algebraic calculus is offered to manipulate these equations. Circuits in the system can be either closed circuit, with zero impedance to current flow, represented by 0; or open, with infinite impedance, represented by 1. The algebra of the circuits is then defined by appropriate postulates, with + expressing sequential connection, and · expressing parallel connection, reflecting truth tables for conjunction and disjunction. By these postulates and using only induction, a series of easy theorems is proved for both addition and multiplication: commutativity, associativity to the right, distributivity, idempotence relations. Negation is introduced and its behaviour additionally defined. See Figure 8.1 (with X equal to 1 or 0). This algebraic structure is shown to be equivalent to the propositional fragment of Boolean logic (i.e. avoiding considering the interpretation of functions over sets). For the analogies between the two systems, see Figure 8.2.

These operations are thus connected to three basic relay circuits (see Figure 8.3): a relay switch between two terminals a, b is represented in a circuit with the letter X; another relay switch is defined for the plus sign (sequential composition); and one for the multiplication sign (parallel composition).

Now it is possible to formulate the circuit for any function computable by the Boolean algebra above.

$1 \cdot 1 = 1$	$T \wedge T \equiv T$
$0 \cdot 1 = 0$	$F \wedge T \equiv F$
$1 \cdot 0 = 0$	$T \wedge F \equiv F$
$0 \cdot 0 = 0$	$F \wedge F \equiv F$
$1 + 1 = 1$	$T \vee T \equiv T$
$0 + 1 = 1$	$F \wedge T \equiv T$
$1 + 0 = 1$	$T \wedge F \equiv T$
$0 + 0 = 0$	$F \wedge F \equiv F$
$X + \neg X = 1$	$T \vee F \equiv T$
$X \cdot \neg X = 0$	$T \wedge F \equiv F$
$\neg 0 = 1$	$\neg F \equiv T$
$\neg 1 = 0$	$\neg T \equiv F$
$\neg(\neg X) = X$	$\neg(\neg T) \equiv T$

Figure 8.1 Equivalence between Boolean algebra of circuits and truth tables for connectives

[2] Shannon (1938).
[3] Brief explanations of these technologies are given in the following sections of this chapter.

Symbol	Interpretation in Relay Circuits	Interpretation in the Calculus of Propositions
X	The circuit X	The proposition X
0	The circuit is closed	The proposition is false
1	The circuit is open	The proposition is true
$X + Y$	The series connection of circuits X and Y	The proposition which is true if either X or Y is true
$X\,Y$	The parallel connection of circuits X and Y	The proposition which is true if both X and Y are true
X'	The circuit which is open when X is closed and closed when X is open	The contradictory of proposition X
$=$	The circuits open and close simultaneously	Each proposition implies the other

Figure 8.2 Table of analogies between propositional calculus and symbolic relay analysis, Shannon (1938)

$$a \quad \overset{X_{ab}}{\circ\!\!-\!\!\circ} \quad b \qquad \overset{X}{\circ\!\!-\!\!\circ}\,\overset{Y}{\circ\!\!-\!\!\circ} = \overset{(X+Y)}{\circ\!\!-\!\!\circ} \qquad \begin{smallmatrix}X\\Y\end{smallmatrix} = \overset{X\cdot Y}{\circ\!\!-\!\!\circ}$$

Fig. 1   Fig. 2   Fig. 3

Figure 8.3 Basic relay circuits, Shannon (1938)

Example 15 *A function of three arguments $f(x,y,z) = (x+y) \cdot z$ is expressed by the following circuit:*

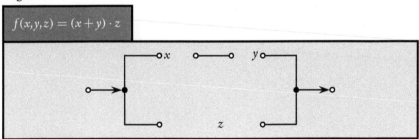

$$f(x,y,z) = (x+y) \cdot z$$

Example 16 *A function of four arguments $f(x,y,z,w) = (x \cdot z) + (-z \cdot w)$ is expressed by the following circuit:*

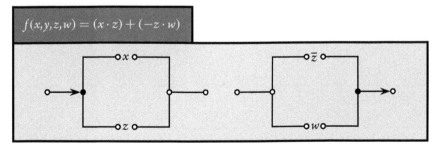

$$f(x,y,z,w) = (x \cdot z) + (-z \cdot w)$$

This model implies further results for the calculus of circuits: De Morgan laws, functions expressed by expansion on variables representing make and break contacts of the circuits to find the valid circuit with the least number of contacts, and simplifications for circuits with the same minimum number of elements. As an example of the reduction process, consider Figure 8.4.

Results obtained for 2-terminal networks are generalized to n-terminal networks and then the theory is shown to be valid for a number of different topologies (like Delta networks, which can be transformed to Wye networks, in turn equivalent to 3-terminal networks by way of associativity of addition). These transformations are also useful to extend results to networks that are not restricted to serial-parallel circuits (e.g. circuits with bridges).

A most important generalization is obtained by distinguishing between independent and dependent variables, where the value of the latter requires computation of the value of the former. Independent variables have hindrance controlled by a source external to the circuit, e.g. those that can be hand-operated by switches. Dependent variables are directly controlled by the network. To compute the set of equations that simultaneously define an entire circuit, the independent variables may be kept fixed, but the dependent ones can change, so that the same network can have different evaluations at different times. For example, a set of equations can be expressed with a variable A from which the evaluation of W, X, and Y depends and a variable B for the evaluation of $C \cdot W$ and $W \cdot X$; see Figure 8.5.

$$W = A + B + C \cdot W$$
$$X = A + B + W \cdot X$$
$$Y = A + C \cdot Y$$
$$Z = E \cdot Z + F$$

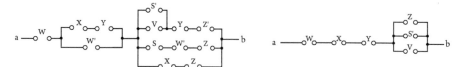

Figure 5. Circuit to be simplified   **Figure 6.** Simplification of figure 5

Figure 8.4 Circuit simplification, Shannon (1938)

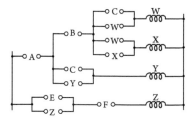

Figure 14. Example of reduction of simultaneous equations

Figure 8.5 Simultaneous circuit simplification, Shannon (1938)

Symbol	In Terms of Operation	In Terms of Nonoperation
X	The switch or relay X is operated	The switch or relay X is not operated
$=$	If	If
X'	The switch or relay X is not operated	The switch or relay X is operated
$'$	Or	And
$+$	And	Or
$(--)'$	The circuit $(--)$ is not closed, or apply De Morgan's Theorem	The circuit $(--)$ is closed, or apply De Morgan's Theorem
$X(t-p)$	X has been operated for at least p seconds	X has been open for at least p seconds

If the dependent variable appears in its own defining function (as in a lock-in circuit) strict adherence to the above leads to confusing sentences. In such cases the following equivalents should be used.

$X = RX + S$	X is operated when R is closed (providing S is closed) and remains so independent of R until S opens	
$X = (R' + X)S'$		X is opened when R is closed (providing S is closed) and remains so independent of R until S opens

Figure 8.6 Operations and their duals, Shannon (1938)

This allows one to simplify products to sums and to closely express negation as electrical impedance, the opposition that the circuit presents to current when voltage is applied: in the case of a dependence of Y from X, $\neg(X \cdot Y) = 0$ means precisely that $(X \cdot Y) = 1$, i.e. $X = 1$ and $Y = 1$. A summary of all operations and their dual is given by Shannon in a Table, reproduced in Figure 8.6

Shannon's seminal work on the logic of circuits is an optimal starting point to reconstruct the evolution of the first generation of computers because of its evident connection with the logical root at the basis of computing machines. It is a known historical fact that Shannon met Turing, first in 1943 when the British mathematician was visiting Bell Labs in New York City, and later in 1950 when Shannon visited England. Turing's visit to Bell Labs lasted around six months, and in that period the two would somtimes have lunch together, discussing both Turing Machines and Information Theory.[4] But it would be historically inaccurate to link any specific significance to these meetings for the origins of computing, as they occured much later than the two contributions of the two pioneers to the foundations, respectively mathematical and engineeristic, of computing. Moreover, Shannon's work is based only on relay circuits, while technological results essential for the so-called first generation of computers include several other important components, like vacuum tubes, drum memory, the stored-program concept, cathode-ray tubes, mercury delay-lines. In the remainder of this chapter, we will explore how the design of these technologies influenced the evolution of the first machines, building on the steps of Shannon's formal work.

---

[4] See Price (1982) for Shannon's recollections of those encounters.

8.2 Early Memories

Relays

Relays are electrically operated switches which were first applied for the electric telegraph in its various incarnations between 1835 and 1840 as invented by Henry, Davy, and Morse; see Figure 8.7.

In these switches, wire is wrapped around an iron core, providing a low reluctance path for magnetic flux and one or more sets of contacts: when the relay is de-energized there is an air gap in the magnetic circuit so that one contact is open, the other closed. When an electric current is passed through the coil, it generates a magnetic field that makes a connection with a fixed contact. When the current to the coil is switched off, the armature is returned by a force, approximately half as strong as the magnetic force, to its relaxed position.

The first computers to work with relays were the Z1-Z2-Z3 machines, developed by Konrad Zuse in Berlin between 1935 and 1941. The most advanced Z3 (see Figure 8.8) was an electrical, binary, 22-bit floating point adder and subtracter. Its input device consisted of a decimal keyboard, output on lamps to display decimal numbers; a control logic allowed it to perform multiplication (by composition of addition) and division (by composition of subtraction). It had 2600 relays, a clock speed of 5Hz, and punched film to store program instructions. It has been proven Turing-complete in 1998.[5]

Vacuum Tubes

Vacuum tubes soon substituted relays in computing machinery. They were invented by Lee De Forest in 1906 and became widely used, despite several limitations including the fact that they were bulky, generating heat, required air conditioning, and were highly

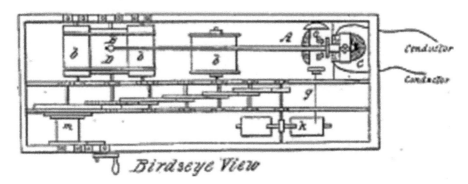

Figure 8.7 Morse telegraph, from Morse (1840) (public domain)

---

[5] See Rojas (1998).

Figure 8.8 A reconstruction of Zuse's Z3. Image courtesy of Computer History Museum

unreliable, as simple air entering the tube could prevent them from working. Vacuum tubes were used to store memory in 1/0 form. They were constituted by three elements: the cathode, the grid, and the plate. The cathode is heated by a filament and it emits electrons; electrons are attracted towards the plate and the grid controls the flow of electrons: by making it negative, the electrons are repelled back to the cathode and the circuit is off; by making it positive, the electrons reach the plate, and the circuit is on; see Figure 8.9.

The first computer to use vacuum tubes was the Atanasoff-Berry Computer, developed at the University of Iowa in 1942, and which was ruled to be 'the first digital computer' in a 1973 court case; see Figure 8.10.[6] The ABC project was started by Atanasoff in the late 1930s and then built with his student Berry between 1939 and 1942. The machine weighted 300kgs, had 1.6km of wiring, and 270 vacuum tubes: 210 for arithmetic unit, 30 for card punching and reading, 30 for charge. The memory was constituted by two drum memory units, holding 50-bit numbers each. The ABC was used to solve linear equations with binary arithmetic and Boolean logic, it was not programmable, and it had no CPU. In terms of speed, it could perform 30 operations per second, store up to 60 numbers per second, for a total of 3000 bits. Its decimal input was stored on punch cards. The drum memory consisted of one or more rows of fixed read-write heads running along the long axis of the drum, one for each track. The machine controller selected the proper head

---

[6] For details of the Honeywell vs. Sperry Rand Court Case, see http://jva.cs.iastate.edu/courtcase.php.

Figure 8.9 Vacuum tubes. Image courtesy of The National Museum of Computing

and waited for the data to appear under it as the drum turned; data was magnetically attached to the drum and the rotational movement from one instruction to the next (called interleaving) was timed to execute series of commands.

Other important machines based on vacuum tubes technologies were the Colossus Mark I and Mark II, in function at Bletchley Park in 1943–44; see Figure 8.11.[7] The Colossus project was started by Tommy Flowers to break the German codes of the Lorenz Machines (this is different from the codes of the Enigma machines broken by Turing's Bomb). The Colossus had 1500 vacuum tubes, 15kW of power, it had a tape transport and photo-electric reading mechanism; a coder and adder; a logic unit that performed Boolean operations; a master control that contained the electronic counters; a printer. It was partially programmable; in terms of performance it was able to process 5000 characters per second via paper tape. Its general functioning consisted in comparing two data streams (encrypted message and internally generated stream) to identify key settings of the Lorenz machines.

The next crucial step in computing machinery was represented by the ENIAC (Electronic Numerical Integrator and Computer), developed at the University of Pennsylvania, USA, between 1946 and 1951; see Figure 8.12.[8] The ENIAC was created by Presper

---

[7] For the whole story of the Colossus and its role in World War II, see Copeland (2010).

[8] For a recent complete analysis of the design, construction, and use of ENIAC, see Haigh et al. (2016). For a more specific view on its programming practices, see Bullynck and De Mol (2010).

Figure 8.10 The Atanasoff-Berry Computer and its drum memory. Images courtesy of Computer History Museum

Eckert and John Mauchly and, crucially for the next chapters of this history, the project had von Neumann as an external advisor. The ENIAC was programmed by physical wiring, needing new wiring at each new program. The machine was over 30m long, over 3m high and 1m deep, and 30 tons heavy. It was used by the US Army's Ballistic Research Lab to compute firing tables. Its design consisted of 18000 vacuum tubes, 150 kw of power, it used the decimal system for programming, could add up to 5000 numbers, do 357 10-digit multiplications or 35 divisions per second. Its programs included loops, branching, and subroutines.

Williams-Kilburn Tube

In the meanwhile, in the UK an alternative system to store data was being developed: the Williams-Kilburn tube; see Figure 8.13. In this system bits were stored on screen

Figure 8.11 The Colossus reconstructed in Bletchley Park. Image courtesy of The National Museum of Computing

Figure 8.12 ENIAC at the Moore School. U.S. Army Photo (public domain)

Aug. 30, 1960   **F. C. WILLIAMS**   2,951,176

APPARATUS FOR STORING TRAINS OF PULSES

Filed Dec. 10, 1947   3 Sheets–Sheet 1

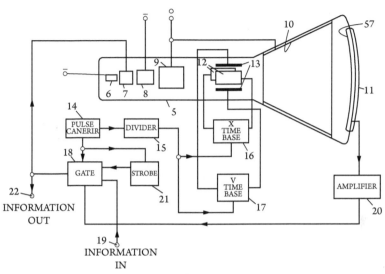

Fig 1.

Figure 8.13 Figure 1 from the 1947 request for US Patent 2951176 for the Williams-Kilburn tube, patented 30 August 1960 (public domain)

as small areas of electronic charge, a metal pick-up plate would detect voltage pulses changes occurring in a regularly timed refresh cycle of 1.2 ms. A signal applied to a position on the screen would generate positive voltage; before the voltage decay happened, the same signal applied to the same position would mean no change was detected and a value 0 stored. If, on the other hand, the position was lightened up with a signal extending the dot with a dash, secondary electrons from the bombardment would be attracted, and the initial positive voltage lost. When that dot position was bombarded again on the next refresh cycle, it would have to acquire a positive potential and that change of charge from the storage area would provide a positive signal at the pick-up plate during the dot period. So a 1 had been stored and detected, and to be preserved the bright-up signal would be extended to cause the dash to be rewritten.

An early computer with cathode-ray tubes and stored program developed in the UK by Williams and Kilburn at the University of Manchester during the period 1948–49 was the Manchester Small Scale Experimental Computer, called the Baby. The Baby is often considered the first properly stored program machine, with a memory of random

A = −S
A = S
A = A−S
A = A+S
S = A
A = A&S
If A<0, CI = CI+1
(i.e. if A negative, skip the next instruction)
CI = S
CI = CI+S
Halt the program

Figure 8.14 4-bits program for the Baby

Figure 8.15 Manchester Mark I. Orphan Work

access of 32 32-bit words stored on tube, it took 1.2ms to execute an instruction, and it had serial binary arithmetic. The memory system included three additional tubes: one for the accumulator, one for present instruction (PI) and its address (CI), and one for

0	00000	/	12	00110	N	24	00011	O
1	10000	E	13	10110	F	25	10011	B
2	01000	@	14	01110	C	26	01011	G
3	11000	A	15	11110	K	27	11011	"
4	00100	:	16	00001	T	28	00111	M
5	10100	S	17	10001	Z	29	10111	X
6	01100	I	18	01001	L	30	01111	V
7	11100	U	19	11001	W	31	11111	£
8	00010	1/2	20	00101	H			
9	10010	D	21	10101	Y			
10	01010	R	22	01101	P			
11	11010	J	23	11101	Q			

Figure 8.16 The Mark I Ecoding

display. The instruction was composed by: a 3-bit function; a 13-bit store address; and 16 bits left unused. Two months after the first version, a 4-bit version was developed (see Figure 8.14) with A standing for Accumulator, S for Store, and $=$ an assignment function.[9]

Mark I, the successor of the Baby, was developed at the University of Manchester in 1949; see Figure 8.15. It had a 40-bit number or two 20-bit instructions (10 for instruction code); memory of random access 32x40 = 1280 bits on two tubes; secondary storage of 2 drums of 32 pages, with one drum revolution in 30ms; 4050 vacuum tubes, 25000w of power consumption; 1.8ms to execute an instruction, and a total of 26 instructions. Programs were entered in binary and the I/O devices included teleprinter, paper tape reader, and display for the tubes. The Mark I had a base 32 programming language, whose encoding was first provided by Alan Turing; it used a ITA2 system to map each character to one of the possible 32 binary values that can be represented in 5 bits (2^5) with some changes; see 8.16.[10]

Its successor, the Ferranti Mark I, was produced by Ferranti Ltd and the University of Manchester in 1951. Its programmers included several famous names: Strachey (important also for work in the semantics of programs, considered in Chapter 7), Turing, and the parents of Tim Barners-Lee. It is one of the first machines to have had audio added for debugging purposes. It had eight tubes for random access memory (each capable of 64 20-bit words = 1280 bits), secondary storage by magnetic drum, revolution in 30ms, 4050 vacuum tubes, an accumulator of 80 bits and included 50 instructions, with 1.2ms instruction time.

[9] See the Programmer's Reference Document, available at http://curation.cs.manchester.ac.uk/computer50/www.computer50.org/mark1/prog98/ssemref.html.

[10] For more technical details on the Manchester machines, see Lavington (1993); the programming handbook of the Mark I written by Turing is available as Turing (1951).

8.3 von Neumann Design

EDVAC

The EDVAC (Electronic Discrete Variable Automatic Computer) was developed by von Neumman, Eckert, and Mauchly at the Ballistic Research Laboratory during the period 1949–61; see Figure 8.17.

Figure 8.17 EDVAC installed in BRL Building 328, from the Archives of the ARL Technical Library. U.S. Army Photo (public domain)

It included 6000 vacuum tubes, 56000w power consumption, and its fame is largely due to the document which von Neumann prepared to illustrate its architecture.[11] It can be characterized by three elements:

1. the stored-program concept,

2. the hardware structure, and

3. the sequentialization of programs with all memory values transferred through the arithmetical unit.

This would later become the common architecture adopted (fully or partially) by most other machines.[12] In his report, von Neumann focuses on the logical control of a *very high automatic digital computing system* defined as a device to carry out instructions to perform highly complex calculations. Instructions reflect the definiteness, clarity, and exhaustiveness required by algorithmic processes. The ability of the device to perform a computation is, right from the start, declared conditional on its ability to overcome malfunctions, the probability of which is not negligible, and so that human intervention is necessary for its resolution, although some automation in error detection is declared possible. This initial remark will be of importance for our analysis in later chapters. The basic elements distinguished by the EDVAC report are today well known:

- the *Central Arithmetical Unit*: it performs basic arithmetic operations $+, -, \times, /, \sqrt{}$; these might be restricted or extended as needed;

- the *Central Control Unit*: it represents the logical unit, allowing the sequencing of the program operations by indicating the address of the next instruction, fetched from memory; this unit needs to be *all purpose*, i.e. independent from the specific instructions stored in memory;

- the *Memory*: it stores all program instructions and data; program instructions are binary values; the Central Control Unit decodes those values to choose the appropriate next instruction; vacuum tube elements are suggested to play this role;

- the *Recording Unit* constituting the *Input and Output Device*: it permits the transfer of information from the Recording Unit to the Central Arithmetical Unit, Control Units and Memory, and back; note that data in the Recording Unit is likely to be present in decimal, hence a conversion to binary is required, for which the Central Arithmetic Unit is used.

The part of memory required for carrying out the operations can be moved to the input, using stacks of cards or tape to this aim, but speed requires essentially relying on the memory unit. One guiding principle for the hardware construction and subsequently for its coding is the following:

---

[11] von Neumann (1945).

[12] This is not to say that other designs were not experimented with, for example in the parallel architecture developed by the Whirlwind I and II since World War II. A critical analysis of the von Neumann architecture was presented in Backus (1978).

> The device should be as simple as possible, that is, contain as few elements as possible. This can be achieved by never performing two operations simultaneously, if this would cause a significant increase in the number of elements required.[13]

In other words, inputs are presented sequentially (i.e. simultaneous signals are avoided) and are taken to be of a length up to 30 binary digits, represented by a line emitting 30 signals (on/off) on an equivalent number of successive periods in time. The draft document proceeds with illustrating the construction of circuits for elementary arithmetic operations,[14] followed by the composition of several circuits through transfer of data from one input to another, to the arithmetic unit and to the output as by function composition:[15]

1. i, j are used to denote respectively two functions in which the input of two distinct memory parts I_{ca}, J_{ca} is passed one to another and then to the output; these operations can be further enhanced with feedback of the output to the input;

2. branching, denoted as operation s: conditional execution is obtained by expressing a function of the form 'if $x \geq y$ then u, else v' which requires four variables broken up into two variable operations: first, check the condition '$x \geq y$ or $x < y$' (through subtraction) and save the value; then, load u and v and compare to the result of subtraction; choose accordingly;

3. decimal-to-binary conversion db;

4. binary-to-decimal conversion bd.

Together with the operations for $+, -, \times, /, \sqrt{}$, this amounts to a total of 10 types of commands or orders. Programming the EDVAC was low-level encoded and relied on distinguishing code from data. The former is characterized by four categories of orders:

1. orders to instruct the Control Unit to execute one of the 10 orders illustrated above;

2. orders to instruct the Control Unit to transfer a value from a memory location to another, or from memory to the Arithmetical Unit, or within the latter; all these types can be replaced just by operations of re-routing of transfer always through the Arithmetical Unit (transfers from Memory to Arithmetical Unit and from the latter to the former);

3. orders to instruct the Control Unit to transfer itself from a memory location to another to read a new instruction; the normal case is for the Control Unit to read orders in sequence waiting in the appropriate delay organ, but there must be orders which occasionally require the Control Unit to switch to specified locations in memory, hence corresponding to *jump instructions*. Such instructions allow so-called *minor cycles* to be entered, after which the Control Unit returns to the main sequence of orders: this is called a *transient transfer*, the logical equivalent of a

---

[13] See (von Neumann, 1945, sec.5.6).
[14] See (von Neumann, 1945, secs.7.0–10.0).
[15] See (von Neumann, 1945, secs.11.1–4).

subroutine, for which the initial memory location of the minor cycle must be remembered for the major cycle to come back to its successor; otherwise, the transition can represent the switch to a different sequence, which is then followed as the natural sequence of orders: this is called a *permanent transfer*, the logical equivalent of a proper GOTO instruction;

4. orders to control the I/O devices.

Each minor cycle is composed of 32 memory units. The first memory unit i_0 distinguishes orders (with $i_0 = 1$) from standard numbers (with $i_0 = 0$). The digit farthest left indicates the sign of the number (with $i_{31} = 0$ designating $+$ and $i_{31} = 1$ designating $-$). Each of the order types listed above has then an encoding in the appropriate remaining digits.[16] To increase efficiency, several orders not requiring the full length of a minor cycle can be composed in one; this nonetheless must be limited when

(i) it would lead to simultaneous execution of several operations (in contrast to the above-mentioned principle of strict sequencing);

(ii) it disturbs timing, and

(iii) it is only meant to free memory, rather than to simplify the logical structure of the code (which is always to be preferred).

In turn, specific order types will not be combined. Finally, the code is offered in terms of all minor cycles (standard numbers and orders) according to four elements:

1. type (either a standard number, an order, or a combination of orders);

2. meaning (the function executed);

3. short symbol (a natural language or symbolic description);

4. code symbol (the 32 binary digits description).

Let us consider some basic programs as examples.[17]

Example 17 *The basic high-level structure of a program to perform addition can be formulated as follows:*

```
// X + Y = Z
LOD X
ADD Y
STO Z
HLT
X := 2
Y := 2
```

---

[16] See (von Neumann, 1945, sec.15.3).

[17] These examples are adapted from the instruction set of the simulator for the von Neumann architecture freely available at vnsimulator.altervista.org.

The program starts by loading through the data bus the value 2 of the variable X in the Instruction Register, passing it to the Arithmetic Logic Unit via the Decoder and then storing it in the Accumulator; then the instruction ADD is loaded in the Instruction Register, performed on the value 2 of the variable Y and calling the stored value from the Accumulator; the operation and its arguments are loaded in the ALU via the Decoder and the result 4 stored back in the accumulator; finally, the next instruction calls the value from the accumulator and stores it in the variable Z via the Addresses bus; finally the halting instruction is loaded in the Instruction Register to halt the program. For the given values, the Program Counter counts five operations.

Example 18 *The basic high-level structure of a program to compute an expression $X > Y$ is given as follows:*

```
// X > Y ? Z
LOD X
JMZ 11
SUB #1
STO X
LOD Y
JMZ 15
SUB #1
STO Y
JMP 1
// X < Y
LOD #0
STO Z
HLT
// X > Y
LOD #1
STO Z
HLT
X := 3
Y := 2
```

The program starts by loading through the data bus the value 3 of the variable X in the Instruction Register, passing it to the Arithmetic Logic Unit via the Decoder and then storing it in the Accumulator; the second instruction is a JUMP to memory cell Z to check whether there is already a value 0 to load as a result (as declared at instruction 11); if not, then the instruction SUB is loaded in the Instruction Register, performed on the value 1 and calling the stored value from the Accumulator; the operation and its arguments are loaded in the ALU via the Decoder and the result 2 stored back in the accumulator; the next instruction is loaded in the Register, it calls the value in the Accumulator and stores it in the variable X; now the program loads through the data bus the value 2 of the variable Y in the Instruction Register, passing it to the Arithmetic Logic Unit via the Decoder and then storing it in the Accumulator; again, a JUMP to cell Z is performed to check whether there is already a value

1 to load as a result (as declared at instruction 15); if not, then the instruction SUB is loaded in the Instruction Register, performed on the value 1 and calling the stored value from the Accumulator; the operation and its arguments are loaded in the ALU via the Decoder and the result 1 stored back in the accumulator; the next instruction is loaded in the Register, it calls the value in the Accumulator and stores it in the variable Y; now the program jumps back to the instruction at line 1 and the cycle is repeated (loop). The loop is executed until one of two conditions is obtained: if the instruction at line 11 is the first condition to be satisfied, it means the program has exhausted the variable X before having exhausted the variable Y, hence $X < Y$ and the value 0 (as a 'false') is stored in cell Z and the program halted; if, instead, the instruction at line 15 is the first condition to be satisfied, it means the program has exhausted the variable Y before having exhausted the variable X, hence $X > Y$ and the value 1 (as a 'true') is stored in cell Z and the program halted. For the given values, the Program Counter counts 18 operations.

EDSAC and UNIVAC

The EDSAC (Electronic Delay Storage Automatic Calculator, see Figure 8.18) was developed by Maurice Wilkes and his team at the Mathematical Laboratory of the University of Cambridge during the years 1949–58. It was explicitly based on the EDVAC design, with mercury delay lines for memory (256 36-bit words up to 1024-word), vacuum tubes for logic, punched tape and teleprinter for I/O. The instructions included: add, subtract, multiply, collate, shift left, shift right, load multiplier register, store/-clear accumulator, conditional skip, read input tape, print character, round accumulator, no-op and stop. The initial orders were wired on switches and loaded into memory; since 1949, an assembler was used to relocate memory, arrays were accessed by directly altering memory locations and subroutines were in use in programs.

The mentioned mercury-delay lines (see Figure 8.19) were a novel introduction, based on radar technology: data was compressed to sound waves and sent through the tube filled with mercury to slow them down; delay was calculated as to get the pulse to the receiver at the time the instruction was expecting to read it; the right pulse was found by comparison of timing of multiple pulses with a clock, the pulse was kept apart from the tube surface and reflections avoided; moreover, the tube needed to be kept at a fixed temperature. One machine equipped with such a storage device was the UNIVAC I (Universal Automatic Computer), the next machine developed by Eckert-Mauchly Computer Corp./Remington Rand, in 1951. It was a decimal-based machine, with 5200 vacuum tubes, weighed 13200kg, capable of producing 125kw of power at a speed of 1905 operations per second, and a 2.25MHz clock. Its memory consisted of 1000 words of 12char (11 decimal digit plus sign) in 10 channels of 100 mercury delay registers. The I/O devices were buffers of 60 words each in 12 channels of 10-word mercury delay registers, tape drives and typewriter, and printer to read tape and print. The available instructions consisted of six alphanumerical characters, digits represented internally in binary.

Figure 8.18 EDSAC I, June 1948. Copyright Computer Laboratory, University of Cambridge, Reproduced by permission

Figure 8.19 The UNIVAC Mercury Delay Tubes. Image courtesy of the Computer History Museum

8.4 Universality and All-purposefulness

We have identified two elements that characterize the rise of the first generation of computing machines:

1. the common architecture at the basis of the EDVAC and then of the EDSAC, known in the literature as the von Neumann architecture;

2. the stored program concept.

The importance of the latter is also indicated by an intense historical and philosophical debate concerning the origins of this concept, and about which artefacts can actually be considered implementing it. The debate concerns mainly the relation between the technical solution for memory usage and the theoretical principle of universality expressed by Turing's generalization of mechanical computation. It seems intuitively clear, as we have illustrated in Section 5.2, that Turing's idea of locating the instructions of the program to be read by the Universal Machine on its tape, alongside the input to be given to such a program, is the theoretical basis for the stored program concept. Nonetheless, it is in general agreed that the report on the EDVAC is the first place in which the idea is formulated explicitly for implementation, by asserting that code (programs) must be stored in order to be executed. Memory is then considered by von Neumann as able to treat parts with different functions, namely code and data. In other words, in the report by von Neumann the abstract principle becomes operative. It remains questionable whether there is any historical evidence of the influence that Turing's work might have had on von Neumann, in particular based on the fact that he read Turing's paper *On Computable Numbers*. This represents just one of the many cases (albeit an important one) in which what we have called the mathematical foundation of computing might have unjustifiably overcome the parallel engineering foundation in later analyses.

It has been pointed out, however, that the first occurrence of the term 'stored-program' dates only to 1949, used by an IBM group for their EDVAC-type machine.[18] The temporal order of historical facts, but more importantly the conceptual order of ideas, have stirred the debate concerning whether the stored program can be assumed to be the same across different conceptualizations of the same computational architecture. The question whether the concept had been implemented *before* von Neumann has been considered in conceptual and historical detail.[19] It seems clear that Zuse's machine Z4 and Aiken's ASCC did not incorporate the stored program concept, they rather used punched cards or tapes to feed the program to the memory and it seems in general untrue that Zuse's computer architectures could be characterized by the principles of the stored program.[20] Besides the disagreement about the attribution of the idea to one or another actor in the history of computing, there seems also to be several levels of conceptual and linguistic differences that have induced further confusion. First of all, it should be noted that the term 'stored program' is a later imposition over several different expressions:

> The reason that the phrase 'stored program' generally did not appear in these documents is simply that the founding fathers tended not to use the word 'program'. Von

---

[18] See (Haigh, 2013, p.243).

[19] In particular, see Copeland and Sommaruga (2015).

[20] See (Copeland and Sommaruga, 2015, sec.5).

Neumann preferred 'orders', as did Zuse (Befehle). Zuse also used the term 'calculating plan' (Rechenplan). Turing, in 1936, introduced the term 'instruction table' for what we would now call a program [...] In the introduction to the first of his Adelphi Lectures (the British counterpart of the Moore School Lectures, although on a much smaller scale), Turing explained that 'the machine will incorporate a large "Memory" for the storage of both data and instructions.[21]

According to this reconstruction, the idea is present already in the mathematical model of Turing, and even the terminology goes back to him and is followed by others like Williams and Kilburn for the Manchester Baby.

Both for those who claim the concept of stored program is well defined since the early steps in the mathematical foundation and for those who, on the other hand, indicate that to be a much later addition of the engineering foundation of computing, it seems that a more analytic presentation is required. First of all, it is clear that the 'stored program concept' should not be identified with the entire von Neuman architecture. Instead, a precise list of components for the latter should be devised:[22]

1. *the modern code paradigm*: this term refers to program-related elements in the von Neumann report; while certain aspects are preserved by later designs (automatic execution, sequential structure, association of each instruction to an atomic operation, predetermined execution sequence, presence of jumps, dynamically changeable addresses for instructions), many differences can be identified, including memory size and characteristics, instruction format, addressing modes, and treatment of code modification;

2. *the von Neumann architectural paradigm*: this refers to the basic structure of organs found in the von Neumann report, including the separation of memory from control and arithmetic;

3. *the EDVAC hardware paradigm*: this refers to the use of a small number of components in a specific composition which anticipated several other attempts by companies like IBM and Bell Labs.

Our analysis in the previous sections of this chapter has highlighted the main characteristics of each of these terms. The stored program concept then sits along with these terms, as the feature that allows code to be treated as data by a program. In fact, the remark that several of the early machines like some of Zuse's did not implement the stored program concept is testimony to the fact that this feature might not be considered necessary to attain the generality of a Turing Machine.

But a second point needing clarification is whether the term 'stored program' strictly refers to the intuition in the context of the von Neumann report, or rather can be qualified by a different and in part complementary analysis, requiring a distinction between types

---

[21] (Copeland and Sommaruga, 2015, p.76).
[22] This list is offered in Haigh et al. (2014).

of memory. This was first suggested by Eckert,[23] while an additional feature is the requirement that memories be differently manipulated by programming.[24] A full analysis in this direction provides the distinction of the layered programming paradigms associated with the stored program concept, thus formulating a different clarification of the concept:[25]

1. *paradigm P1*: programming consists of physically rewiring the machine;

2. *paradigm P2*: programming is facilitated by encoding instructions and storing them in memory (e.g. punched cards);

3. *paradigm P3*: encoded instructions are stored in read-only memory, so that they can be read repeatedly;

4. *paradigm P4*: encoded instructions are stored in read memory of the same kind as that used for execution;

5. *paradigm P5*: encoded instructions are stored in read-write memory feasible for editing them via insertion, manipulation, or deletion;

6. *paradigm P6*: editable encoded instructions can be those of the running program or those of a different program stored in memory.

This analysis illustrates, again, the need to apply different levels of abstraction to computing concepts in order to define them properly. The stored program concept identifies, in particular, a basic tension with the notion of *universality*.

From the mathematical, abstract viewpoint, universality refers to the dependence of the computational behaviour from an algorithmic encoding, while preserving its independence from any specific linguistic encoding.[26] In this sense, that the Universal Turing Machine is able to simulate any individual machine renders the universal feature of computability as the ability to allow different equivalent encodings for the same algorithmic structure. We define it as follows:

Principle 7 (Universality) *A machine is universal if it can simulate any computation.*

From an engineering viewpoint, instead, one has to account for the physical implementation that makes such simulation of computation realizable, and for this the term *all-purposefulness* can be used. In this second sense, one has to start considering the different elements at hardware and software levels that have (progressively) made such generality possible. In particular, an all-purpose machine requires the distinction drawn by von Neumann between instructions needed by the problem at hand (software issue) and their storing and accessibility through appropriate physical organs (hardware issue). While both elements can be associated with the UTM, in particular the former with the fetch-execute program inscribed on its tape and the latter with the program of the

---

[23] See Eckert (1947).

[24] See (Randell, 1994, p.13).

[25] This analysis is offered in (Copeland and Sommaruga, 2015, sec.4).

[26] Cf. Principles 5 and 6 in Chapter 7.

Universal Turing Machine,[27] it is also clear that this reduction makes the hardware–software distinction disappear. Hence, a counterpart principle in terms of the physical instantiation of the Universal Machine can be offered:

Principle 8 (All-purposefulness) *A machine is all-purpose if its architecture allows for the capacity to store, access, execute, and modify code (including its own).*

The notion *all-purposefulness* can be logically reduced to that of *universality*, and this is an important feature of the foundations of computing we are investigating. Nonetheless, taking into account both abstract and physical properties of computing allows one to clarify several aspects pertaining to each. In the following chapters, we will assume that the presented analysis is for all-purpose machines and we will consider more closely how the hardware and software evolution have reflected problems that could not be formulated at the abstract level of universality.[28]

Exercises

Exercise 81 *Explain in general terms the difference in how they work between relays, vacuum tubes, and mercury-delay lines in the development of early computers.*

Exercise 82 *Explain how a vacuum tube can represent a binary digit.*

Exercise 83 *Which was the first computer to run under the stored-program architecture? Which one in the UK?*

Exercise 84 *Evaluate the two circuits presented in Example 15 and Example 16 of the first section of this chapter for some assignment of the variables to values $1, 0$.*

Exercise 85 *Design the circuit for the following formula: $f(x, y, z, w) = w \cdot (x + (y \cdot (z + \neg x)))$. Then evaluate the circuit for the following assignments:*

- *the circuit with output w is open;*
- *the circuit with output z is closed;*
- *the circuit with output y is closed;*
- *the circuit with output x is open.*

Exercise 86 *Describe in some detail the components and functioning of the von Neumann architecture.*

Exercise 87 *What is the total number and what are the operations of the EDVAC?*

---

[27] See (Rapaport, 2018, p.368).

[28] Note that this analysis leads directly to the philosophical question: 'what is a computer?', and to its variations, including pancomputationalism and the question whether the brain is a computer. This debate is beyond the scope of this volume, and we shall not explore it. For an extensive overview of contributions to this debate, see Rapaport (2018).

Exercise 88 *Go the Von Neuman Architecture Simulator at http://vnsimulator.altervista. org/index.php. Load the file for subtraction.*

- *Set the delay time to 2500ms; this will allow you to see the operations performed by the architecture in a slow time frame*
- *Set the program counter to +1; this is the number of instructions performed at each step*
- *Load two values, for variables X and Y*
- *Hit the Step button to see the first instruction; when the machine halts, it has finished the first operation, you will need to hit Step again to perform the second one and so on*
- *Complete the program execution.*

   Run a few instances of this program with different values and write down in detail the functioning as you see it working.

Exercise 89 *Do the same as above with the file* number_is_even.vnsp.

Exercise 90 *Explain the different interpretations of the stored program concept illustrated in this chapter.*

Exercise 91 *Explain the difference in levels of abstraction between the notions of universality and all-purposefulness illustrated in this chapter.*

9 The Laws of Evolution

Summary

In Chapter 8, we recalled the birth of the first generation of computers in their technical aspects in order to characterize them by the concepts of universality and all-purposefulness. In this chapter, we explore the technologies that have made possible the rapid development of computing machinery, essential for exploring the meaning of the laws of hardware and software evolution.

9.1 Computing grows

The first generation of computers is linked to the invention of vacuum tubes as the main technology for volatile memory, and the definition of the architecture to relate it to arithmetic and logic units, and input/output devices. All the following generations of computers can similarly be linked to new technologies: transistors and magnetic core for the second generation; integrated circuits for the third; miniaturization and parallelism for the fourth. This evolution also indicates the progressive improvement of computing technologies through non-volatile memory, size and reliability, efficiency. These concerns were not just induced by the creation of new technologies but pushed by the extension of use of computing from purely scientific purposes to business, industrial, and safety-critical applications. These new principles guided the development of computing, from hardware to programming languages and software systems.

As in Chapter 8, our main concern is not to provide a full historical overview of the many machines developed, nor a detailed illustration of the technologies that have defined the various generations of computers, either at hardware or at software level. Rather, we will use some exemplificatory cases in order to provide the reader with an idea of the speed and evolution that the engineering foundation has seen in the decades

On the Foundations of Computing. Giuseppe Primiero, Oxford University Press (2020). © Giuseppe Primiero.
DOI: 10.1093/oso/9780198835646.001.0001

between the 1950s and the 1990s. Our ambition is to highlight some of the important principles characterizing technology to better understand the conceptual evolution of computing.

9.2 New Memories

Transistors

Relays and tubes were spreading across different technologies, not only in computing but in telephony as well. Their main limitation was related to their short life and high power consumption. The revolution that made their replacement possible was initiated at Bell Labs by the invention of semiconductor material and transitors: the idea was to create a device where electrons would travel short distances within a solid environment, without filaments, moving parts, or vacuum.[1] Transistors were developed from 1946, in the group led by William Shockley and with the main contribution of John Baarden and Walter Brattan, who filed their patent application in 1948 (patented 1950):

> This invention relates to a novel method of and means for translating electrical variations for such purposes as amplification, wave generation and the like. The principal object of the invention is to amplify or otherwise translate electric signals or variations by use of compact, simple and rugged apparatus of novel type. Another object is to provide a circuit element for use as an amplifier or the like which does not require a heated thermionic cathode for its operation [...] A related object is to provide such a circuit element which requires no evacuated or gas-filled envelope.[2]

The underlying physics consisted of the known concepts of band gaps and of two types of conduction (n-type for negative carriers and p-type for positive carriers). The group at Bell Labs focused on crystals of silicon and germanium and on the field-effect mechanism, according to which an electric field applied to a semiconductor could modify the density of the body and thus its conductivity, a problem already investigated by Shockley.[3] The problem presented by immobile charges in the surface of the semiconductor which would terminate the electric field was solved by the transitor effect:

> When the two point contacts are placed close together on the surface and direct current bias potentials are applied, there is a mutual influence which makes it possible to use the device to amplify alternate current signals.[4]

---

[1] See (Ross, 1998, p.8).
[2] (Baarden and Brattain, 1948a, p.1). For extensive analyses of the technical and historical context of this technology, see Hornbeck (1985); Riordan and Hoddeson (1997); Teal (1976).
[3] See (Ross, 1998, p.9).
[4] (Baarden and Brattain, 1948b, p.230).

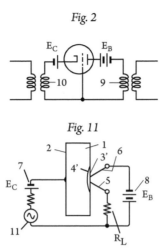

Fig. 2

Fig. 11

Figure 9.1 Schemes of a transistor, (Baarden and Brattain, 1948a, Sheet 1 of 4)

The Bell Labs contributed again with the crucial improvement offered by the Metal Oxide Semiconductor Field Effect Transistor invented by Dawon Kahng and John Atalla. The so-called MOS transistor consists of a three-terminal, solid state electronic device which controls electric current passing between two of the terminals by applying voltage to the third one; see Figure 9.2:

- the *base* is the gate controller device for the larger electrical supply;
- the *collector* is the larger electrical supply;
- the *emitter* is the outlet for that supply.

The amount of current flowing through the gate from the collector may be regulated; the transistor is being used as a switch with a binary function: five volts flowing means the device is ON, less than five volts means the device is turned OFF. The engineering phase to actually build the transistor lasted around eight years, punctuated by hurdles and solutions.[5] Initially two types were developed, point-contact and alloy junction transistor. The latter improved efficiency and ease of use, simplifying automated production. From the early 1950s, silicon was identified as a better material for transistors and by 1954 the first manufacturable silicon transistor was developed at Texas Instruments.

The use of the transistor in logic applications was successful for its small size and requirement for low power, which made it possible to pack many of them in a small space without excessive heat generation and with low propagation delays.[6] An early machine to make use of this new technology was the Manchester Transistor Computer, developed as a prototype by the University of Manchester in 1953, and then commissioned in 1955;

[5] See (Ross, 1998, p.15).
[6] See (Ross, 1998, p.18).

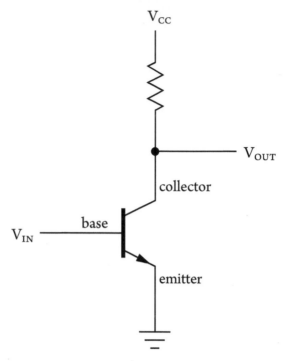

Figure 9.2 Circuit diagram of a transistor. CC BY-SA 3.0

see Figure 9.3. This final version had a total of 250 point-contact transistors, 1300 point diodes, resulting in 150 watts power consumption. It was capable of 1,5 hours error free (at hardware level) run, it contained a small number of tubes in its clock generator, was able to add two 44-bit numbers in 1.5 drum revolutions with the drum operating at 3000rpm; a division subroutine took one second, a square root operation took 1.3 seconds.

The ability to increase the number of transistors induced an increase in reliability and computational power, as well as a loss in power consumption. The TRADIC (TRAnsistor DIgital Computer), for example, built by J.H. Felker and L.C. Brown at Bell Telephone Laboratories, during the period 1951–54, produced 30 watts for 1MHz clock (which as for the Manchester machine had a vacuum tube) but consumed less than 100 watts of power, with 16-digit word length, performing multiplication or division in less than 300 microseconds and addition or subtraction in 16 microseconds. The memory was provided by 18 electric delay lines, each with 1 word capacity; it had provision for 13 16-digit constants stored in toggle switches; programs were stored in a plug board.

This constant evolution in computational abilities, associated with reduction of power loss, became more evident in shorter periods of time. The other aspect significantly improved by the introduction of transistors, and which would effectively require a trade-off with computational speed, was reliability. The Harwell Cadet (Transistor Electronic Digital Computer), developed by E.H. Cooke-Yarborough in 1955–56 (see Figure 9.4), had a low clock speed of only 58kHz; accordingly it was a slower machine, but it was able

Figure 9.3 The Manchester Transistor Computer. Orphan Work

to add up to eight numbers concurrently. It included 324 point-contact transistors, 76 junction transistors for data read from the drum and it performed well with computing runs of more than 80 hours without errors. Hence, a balance between computational power and reliability was becoming an important issue that would be stressed in the later decades by novel applications of computing technologies.

In fact, one important passage in the evolution of computing machines was the expansion to the business industry.[7] The impact of business computing is largely acknowledged in its economic and social aspects, for example for how it changed productivity and how it significantly shifted gender perceptions of computer users.[8] But it also had a major effect on the design and analysis of problems related to programming and implementation, including aspects of correctness, thus partly modifying the meaning of this mathematical notion. The machine generally acknowledged for an early use of computing technology in a business context is the LEO I (Lyons Electronic Office I), developed between 1949 and 1951. With its successor LEO II, it was a variation of the EDSAC design based on

---

[7] For extensive case study analyses in the history of business computing, see e.g. Hevnes and Berndt (2000); Land (2000); Haigh (2001); Mowery (2003). For an overview of the most significant early commercial computers in the US, see (Ceruzzi, 2003, Chapter 1). For a history of business computing in the United Kingdom focusing on ICL, see Campbell-Kelly (1989).

[8] See for example Hicks (2017).

Figure 9.4 The Harwell Cadet, 1955. Photo by Michael Wilson, Public Domain

the von Neumann architecture: LEO I had twice the memory size of the EDSAC of 2048 bytes, 17-bit words; 228 separate electronic units, almost 6000 vacuum tubes. The transition from scientific to business applications is of major importance for the issue of correctness: programs for business tended to be much bigger, due to the economic need of carrying out as many operations as possible once the data was loaded, and because of their less logical structure compared to scientific applications, often due to regulatory and commercial variations. This difference impacted runtime execution: jobs were time critical and had to be completed to a rigorous time schedule despite slow I/O mechanisms taking up to 90% of the entire running time. Hence, the idea of showing a program correct slowly turned to that of making a program efficient: in developing the LEO, for example, a large body of techniques were devised to avoid mistakes in programming and to guarantee data correctness, with the aim of securing applications integrity and reduce the risk of having to re-run the programs. These techniques included flowcharting, understandable documentation, redundancy in data, and program checking, both manual and automatic. Moreover, systematic hardware testing was developed in these early phases of business computing. From a methodological viewpoint, systematic fault diagnosis was executed to discover the causes of *malfunctioning*.[9]

---

[9] For a fully documented analysis of validity and correctness techniques in LEO I and II, see Arif et al. (2018). Malfunction limitation is an important topic that will guide the engineering foundation throughout the following decades and which we will explicitly address in its conceptual relevance in Chapter 11.

Figure 9.5 LEO I control desk. Credits: LEO Computer Society

In the decades to come, this association between reliability and efficiency as appropriate complements to formal correctness were further strengthened within business computing by its strongest representative: IBM.[10] The company was since its early days leading the development and use of new technologies and already with its IBM 608 between 1957 and 1959, often considered the first commercially available machine on the market, included magnetic core memory. The machine could be programmed through a control panel with a 40 nine-digit numbers memory, 18 digit accumulator; it was capable of 4500 additions per second, the maximum time to perform multiplication was 19.8ms, and 22.4ms for division. Its programs could be as long as 80 (non-sequential) program steps. Its successor, the IBM 7070 added a card for integrated transistors and also used magnetic core to contain the stored program and all data used in its operations. Data read from any of the input-output or storage units was brought to core storage and back. It contained 9990 words, with a word consisting of 10 numerical digits and a sign (plus, minus, or alpha). Each word in core storage was addressable and each digit was represented by a combination of two bits out of possible five. The total number of possible combinations was ten—one for each numerical digit. An alpha-numerical word in core storage contained five characters, each represented by two digits, with characters in two-digit code sequence. These two early examples show how IBM was ahead in the development of second and third generation computers, in which random access

---

[10] See also Chapter 10 for an analysis of data validation techniques stemming from IBM research.

memory (RAM) was predominantly based on magnetic cores and processing was based on integrated circuits.

Magnetic Core

In a paper from 1953, Eckert refers to magnetic cells as follows:

> Of the more speculative and experimental memory systems, the most promising are those utilizing materials whose magnetic properties are characterized by square hysteresis loops.[11]

The hysteresis loop of a ferromagnetic material is the relationship between the induced magnetic flux density and the magnetizing force. The system of toroidal cores of magnetic material was introduced in a patent application from 1952; see also Figure 9.6:

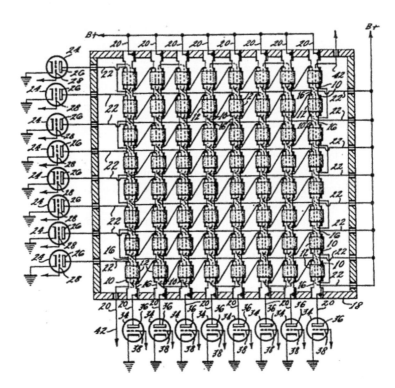

Figure 9.6 Scheme of magentic core, (Rajcham, 1957, Sheet 1 of 2)

---

[11] (Eckert, 1998, p.23).

toroidal cores of magnetic material [...] may be arranged in a two-dimensional storage array [...] in columns and rows, have a separate coil inductively coupled to all the cores in a column.[12]

Writing consisted in selecting both row and column coil and charging them with current in the same direction, causing the cell common to the two wires to flip from one polarity to the other. To read the condition of any core there was a reading coil which listened for changes in its magnetic shift: if the saturation of the core was the same as the current polarity of the reading coil, no output was obtained; a pulse was induced by a polarity opposite to that of the reading driving currents. The main features of these memories were:[13]

- *non-volatility*: no memory loss on power failure;
- *power efficiency*: little power required for reading or writing, none for regeneration;
- *system efficiency*: no drift in memory location (known as the resolution problem).

Along with business, another area which would be crucially impacted by the introduction of computing technologies and for which reliability and efficiency were of the essence was advanced safety critical engineering, and in particular the space race of the 1960s. The Apollo guidance system, for example, in its final design had 36k of fixed memory (core-rope ROM for programs) and 2k of erasable memory (coincident-current ferrite cores for calculations' results, data and logic operations). This resulted from several design iterations, as limitations on memory affected significantly software development, both in terms of program size and addressing issues.[14]

9.3 Miniaturization, Parallelism, and Compatibility

The development of the third and fourth generations of computers was thus led by an increasing attention towards reliability and efficiency, with a huge impact on the design and production of computing machinery, an issue which can be analysed along three main axes: miniaturization, parallelism, and compatibility.

Miniaturization would be the driving force towards personal computers, but it would also facilitate the convergence of mainframe and minicomputing. Initially miniaturization was prompted by the need to overcome problems with the mass production of transistors, in particular related to speed, dimensions, and reliability. Well-functioning was negatively affected by signal propagation delays and this led to the reduction of space usage through miniaturization of the components. This, in turn, had an immediate drawback on reliability, addressed by reducing the components' number and simplifying

---

[12] (Rajcham, 1957, p.1).
[13] See (Eckert, 1998, p.24).
[14] See (Tomayko, 1988, ch.2).

their connections. Reliability was—for obvious reasons—the main requirement for the American Air Force, which had been supporting extensive research on the problem of interconnecting components since the 1950s. Cost reduction as well as functionality by error limitation were mostly the concerns of civilian electronics. Integrated circuits were the resulting next technical advance.[15]

Precursors of this new technology were Werner Jacobi at Siemens in 1949 and Geoffrey Dummer of the British Ministry of Defence in 1956. In 1958–59, integrated circuits were invented independently by Jack Kilby at Texas Instruments and Robert Noyce at Fairchild Semiconductor. The semiconductor material used to integrate the circuits was usually silicon and it guaranteed lower cost and better performance. According to Kilby's Patent:

> It is therefore a principal object of this invention to provide a novel miniaturized electronic circuit fabricated from a body of semiconductor material containing a diffused p-n junction wherein all components of the electronic circuit are completely integrated into the body of semiconductor material. It is further object of this invention to provide a unique miniaturized electronic circuit fabricated as described whereby the resulting electronic circuit will be substantially smaller, more compact, and simpler than circuit packages heretofore developed using known techniques.[16]

Thus, despite the push for increasing reliability mainly coming from the military sector, the invention which would guarantee its realization came from private research. The military was, nonetheless, the first customer for the new product. In the already mentioned Apollo guidance system computer, integrated circuits were introduced with the Fairchild design which guaranteed simplicity of the logic and compatibility of the components. Price drop was also a significant consequence of the large use of the new technology.

The other market which obviously gained enormously by the miniaturization, efficiency, and economic savings allowed by the integrated circuit was that of minicomputers. Standardization of chip connection, realized by Globe-Union, made it possible for producers to design a single circuit board that would be compatible with products by several computer manufacturers. In the early 1960s the small integration scale of integrated circuits allowed up to 10 transistors and 12 logic gates; towards the end of the decade, medium integration scale increased those respectively up to 500 and 100 units. Integrated circuits would soon be put to use for random access memory as well, substituting magnetic core in the ILLIAC IV, which would be groundbreaking also for its impact on parallelism.

After faster and cheaper memory systems came into production, control-and-arithmetic logic units as well as more reliable I/O methods were the main object of research and development. But the essential architecture of machines remained the same.

---

[15] For an overview of the research programs developed by the US Army to increase reliability in circuit production and which led to the invention of integrated circuits, and for a history of the impact of integrated circuits on mainframe and minicomputers, see (Ceruzzi, 2003, Chapter 6).

[16] (Kilby, 2007, p.49).

Fig.7.

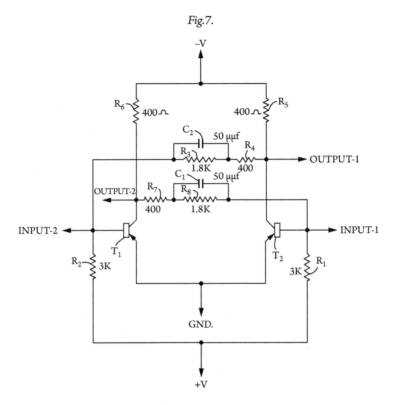

Figure 9.7 Scheme of an IC, (Kilby, 2007, Sheet 3 of 4)

The main limiting factor was still execution speed, for which two main approaches were devised:[17]

- *overlap*: the architecture is modified to allow overlap of operations by two or more components, so that more than one operation is occuring during the same time interval;
- *replication*: components are duplicated several times, e.g. for I/O devices.

A combination of both approaches characterizes the ILLIAC IV search for efficiency and optimization, with a design consisting of four central processing units, allegedly becoming the first parallel computer:

> The limitation imposed by the velocity of light, once thought to be an absolute upper bound on computing power, has been stepped over by several approaches to computer architecture, of which the ILLIAC IV is the most powerful by as much as a factor of four.[18]

[17] See (Barnes et al., 1968, p.370).
[18] See Burroughs (1974).

The ILLIAC IV, designed at the Computer Science Departement of the University of Illinois, used overlapping of buffer and pipeline mechanisms, allowing operations to be broken down into subroutines that could be performed in parallel. Replication up to four times was used by the multiple processor Control Unit, Arithmetic-Logic Unit, and memory. The CU was then re-centralized in the version built by Burroughs Corporation to constrain costs and this had an impact on the structure of a unique CU and more than one ALU, with a large number of both scientific and business problems allowing the design of algorithms performed repetitively on many sets of different data.[19] A higher number of elements in the ALU array (called PE) allowed resolution of progressively more requiring problems: from just large unstructured table lookup to solution of non-linear equations to matrix arithmetic.

The success of the architectural revolution proposed by the ILLIAC IV inventors, in terms of parallelism and the use of integrated circuits for random access memory, was at the time only partial, but it kickstarted a whole new programming paradigm which today is central. Moroever, it was the sign of another important change to come. The immense impact of the miniaturization process on cost reduction and power expansion of computing technologies is best seen through the invention of microprocessors: thousands of integrated circuits on a single silicon chip, serving from CPU and memory to input-output controls. The invention of the microprocessor has been era-defining:

> in less than 30 years, a single chip weighing less than one gram, occupying a volume smaller than a cubic centimeter, dissipating less than one Watt, and selling for less than ten dollars could do more information processing than the UNIVAC I, the first commercial computer, which used 5200 vacuum tubes, dissipated 125 kW, weighed 13 metric tons, occupied more than 35 m 2 of space, and sold for more than one million dollars per unit.[20]

The invention is strongly linked with the foundation of Intel by three Fairchild employees: Robert Noyce, Gordon Moore, and Andrew Grove. The inception of the microprocessor was made possible by the already mentioned MOSFET (Metal Oxide Semiconductor Field Effect Transistor) invented by Dawon Kahng and John Atalla at Bell Labs, and the Silicon Gate Technology (SGT, Figure 9.8) invented at Fairchild Semiconductor in 1968 by Federico Faggin and Tom Klein, the latter being central to making the former technology reliable. The self-isolation and surface-effect properties of MOS transistors made it possible to scale them significantly and the silicon gate technology improved their reliability. MOS ICs became five times faster and had twice the number of transistors per unit area than MOS ICs made with aluminum gate (for random logic circuits). RAMs, image sensors, non-volatile memories were also made with the SGT. The first commercially available self-aligned gate MOS IC with SGT was the Fairchild 3708.[21]

---

[19] (Barnes et al., 1968, p.370).
[20] (Faggin, 2015, sec.1).
[21] See Faggin (2009). For a detailed analysis of MOS and SGT see Faggin (2015).

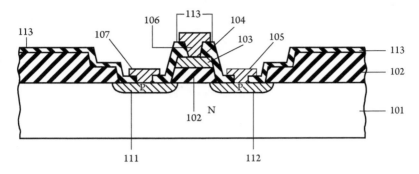

Figure 9.8 Scheme of an MOS device with SGT, Klein and Faggin (1972)

The first microprocessor, the Intel P4004, dates back to 1969: the 4004 was the CPU of a group of four chips that Intel was contracted to deliver to Busicom, a Japanese company manufacturing mechanical and electronic desktop business calculators, selling Mitsubishi mainframe computers, developing software for them, and importing business computers from France.[22] The P4004 was processing 4-bit in 8-bit format (a byte), had 108kHz clock speed, 2300 transistors, up to 1kB of program memory and 4kB of data memory and it qualifies as the first large-scale integration chip.

From that first chip, the history of microprocessors has been a well-known saga of increasing efficiency. In the 1970s, large-scale integration meant up to 20.000 transistors and 10.000 logic gates. In the 1980s, very large-scale integration increased transistors respectively to 1.000.000 and logic gates to 100.000. From the mid-1980s, ultra-large-scale integration completed the size explosion to over one million transistors and over 100k logic gates per circuit. In 2005 the first dual core microprocessor was produced, the Athlon AMD 6400+: shared dual channel and control logic; ability to multi-thread; 256, 512, or 1024kB full speed, per core; 2000–2400 (up to 3100) MHz clockrate. In 2014 the Power8 was produced, with 96 threads on 12-core chip; 64kB data and 32kB instruction caches; 96MB of cache; 1TB of RAM; 2.5 to 5gHz clockrate. As of 2018, the two major producers have the following top-of-the-line microprocessors:

- AMD Ryzen 7 2700X (2nd generation): a microprocessor with 8 CPU cores, allowing 16 threads, a base clock speed of 3.7GHz, boost clock up to 4.3Ghz, 20MB cache;
- Intel Coffee Lake: a microprocessor with 6 CPU cores, allowing 12 threads, a base clock speed of 3.7GHz, 4.7GHz of maximal Turbo Core Speed, and 12MB of cache.

Along with miniaturization and the crucial role of the microprocessor, compatibility was the other essential component that made the extensive development from mainframe

---

[22] For the background history to the invention of the P4004 and the relations with the Busicom Project, see Aspray (1997) and Faggin (2009).

to mini and personal computers possible. The market for personal computers started towards the end of the 1970s with a full range of products available from the United States, like Tandy's and Commodore. The first commercially available minicomputer is often indicated as the PDP-80, produced by DEC, which only in later models used parallelism and was equipped with medium-scale integration chips. The prices of early mini computers were still close to prohibitive to kickstart the revolution of the home computer. A notable example combining small size and low cost was the Sinclair ZX80, which effectively illustrates the passage to the next generation of computers: the ZX80 was the first computer to be marketed for less than $200 or £100. With 1kB of static RAM and 4kB of ROM, it was programmable in BASIC and had means for *expansion* through an edge connector on the motherboard. Compatibility was a central issue not only at hardware but also at system level. Another example is the choice of 16 bits for the mini's word length led by IBM, substituting the previously existing—and cheaper to produce—6-bit format common among minicomputer manufacturers. Already in 1964, IBM announced the choice of the byte standard of 8 bits for the encoding of a letter or other character and implementing it as a 32-bit word length in its System/360. Intel followed suit in choosing the same format for its products. Missing compatibility had important effects. Many different standard codes for mini, and later personal computers existed as the result of the missing standardization for the 8 bit of ASCII, which forced IBM to enter the personal computer market with an ASCII machine in the 1980s, abandoning its previous choice for the proprietary EBCDIC code.[23]

This historical overview and conceptual analysis has highlighted three main engineering principles at the basis of the evolution of computing machinery:

1. the effort to reduce space and costs;
2. the need to improve efficiency;
3. the increase in shared design choices.

These aspects all respond to laws of hardware and software evolution, which are analysed in the following sections.

9.4 The First Law

In 1965, Gordon E. Moore, physical chemist and later, as we already saw, co-founder of Fairchild Semiconductor and Intel, opened an article with the following prediction on the state of hardware technology within the following 10 years:

> With unit cost falling as the number of components per circuit rises, by 1975 economics may dictate squeezing as many as 65000 components on a single silicon chip.[24]

---

[23] See (Ceruzzi, 2003, pp.193–4).
[24] (Moore, 1965, p.82).

After analysing the increasing importance of semiconductor devices and silicon for modern electronics, Moore pointed more explicitly to the relation between cost per component in integrated circuits and their number in one single unit:

> the cost per component is nearly inversely proportional to the number of components, the result of the equivalent piece of semiconductor in the equivalent package containing more components. But as components are added, decreased yields more than compensate for the increased complexity, tending to raise the cost per component.[25]

Moore's prediction has been discussed at length, especially during the 1990s, and it is still of crucial importance today.[26] The law can be analysed more precisely as the result of four main factors:[27]

1. *die size*, i.e. the surface area of the silicon wafer;
2. *line dimension*, i.e. the depth, width, and density of circuitry that can be etched on the silicon;
3. *technical cleverness*, composed of device cleverness (i.e. the ability of engineers to design chips using as much of the chip's space as possible) and computer-aided design;
4. *technical innovation*, i.e. the advances in materials and methods for chip production.

These factors must be projected over several distinct time periods, those briefly highlighted above in Section 9.3:

1960–65: Small-Scale Integration;
1965–80: Medium- and Large-Scale Integration;
1980–85: Very Large-Scale Integration;
1985–95: Ultra-Large-Scale Integration.

In the Small-Scale Integration period, the main driver of the law was economics, along with the engineering improvements. The success of integrated circuits was not just a matter of efficiency but also of cost and simplified design, while their functionality would basically be initially the same as that of hand-wired components. In its first formulation, Moore's prediction parametrized reduced costs and the ability to increase components, i.e. density: in simple circuits, the cost per component was already nearly inversely proportional to their number, but with additional component units being added, the cost per chip unit tended to increase. This means that any given technology has a minimum cost with the overall costs falling, a relation that Moore puts in perspective with a graph;

---

[25] (Moore, 1965, p.83).
[26] See e.g. Bondyopadhyay (1998); Schaller (1997); Mack (2011).
[27] See Mollick (2006).

see Figure 9.9. By this inverse relation, the expected cost per component in five years was expected to fall to $1/10^{th}$ of the value in 1965. At the same time, Moore noted that the number of components in integrated circuits had doubled roughly every year since their invention in 1958; see Figure 9.10. This trend was expected to continue well over the decade span predicted by Moore. In this first formulation, therefore, Moore was mainly concerned with the economics of the industry.

Between 1965 and 1968 the slope increase of Moore's Law had a gap. In that period, no semiconductor product reached the potential limit to match the curve (as stated by Moore himself in a later article of 1979). This meant that the prediction was aiming at graphing exclusively the upper potential bound of the products, and this occurred where their production made economic sense. The beginning of the Medium-/

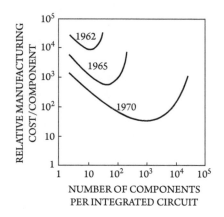

Figure 9.9 Relation between manufacturing costs and number of components, Moore (1965)

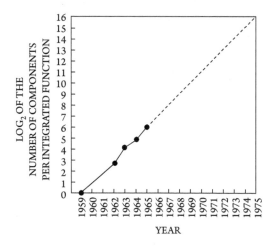

Figure 9.10 Number of components in IC, Moore (1965)

Large-Scale Integration period started with the integration of logic circuits on chips and by the end of this period economics was no longer the main factor in realizing the prediction: technical issues had become more important, as integrated circuits were no longer one among the possibilities, rather the *de facto* standard.

In 1975, Moore gave an updated view on its prediction, taking into account a number of different factors for justifying the expected continuation of the trend:[28]

- *Integrated structures*: the increase of the chip area for maximum complexity, relying mostly on higher density of components on the chip;
- *Circuit and device advances*: improvements in these areas accounted in 1975 for about 100-fold increase to higher density.

In particular, three new uses were found for microelectronics:[29] dynamic random access memory (DRAM) became the main market for semiconductor manufacturers, microprocessors, and universal asynchronous receiver-transmitters (UARTs). The analysis extrapolated over the successive five years gave Moore sufficient ground to expect such a trend to continue in the future, based on the following reasons:

1. the size of the silicon wafer corresponding to one chip (die size) grew principally as a factor of reduced density of defects harming integrated circuits;
2. minimum size device determined by fundamental aspects such as charge on the electron or the atomic structure of matter had not been reached yet.

By the first item, the issue of malfunctioning comes into play again explicitly. The second issue shows how a limit was identified in device and circuit cleverness, which denotes the ability to design circuits in a way that maximizes the topological distribution of units. With circuits approaching the maximum practical density due to physical limitations, the prediction was updated to a doubling of the number of components in integrated circuits every two years (instead of every year). By the time Medium- and Large-Scale Integration were accomplished, the semiconductor industry could rely on both economic factors and technical advancements, in fact the latter influencing the whole economics of the industry. But again a breaking point had been reached.

In the early 1980s, Moore's Law was adapted to predict the densities of dynamic RAM more specifically. But more technologies were being developed, in particular for logic units and memories like CCD (charge-coupled device). It was based on the intended wide availability of CCD units that Moore expected the contribution of circuit cleverness to disappear, hence reducing the slope increasing from doubling every year to every two years. Two main issues are at stake here: first, Moore's new adaptation of his law matched temporally with the transition to the VLSI circuits period; second, he was responding to industry-related progress, directly available to him as part of the very same complex. In

---

[28] See Moore (1975).
[29] See (Mollick, 2006, p.66).

the VLSI period, Moore's prediction was simplified by referring to DRAM chips only, the technology was stable and relatively simple. This meant that the two main factors were line density and die size: from the late 1970s to the mid-1990s, DRAM chip density would increase at a pace of approximately 31% a year, or a doubling every 26 months, thus confirming Moore's prediction. As far as microprocessors were concerned, they would not be measured in terms of transitor-density, but rather in terms of their speed of operation (MIPS—millions of instructions per second). This would still approximate to a 30% improvement per year. The VLSI era therefore expresses Moore's Law in terms of three indicators: DRAM complexity, processor transistor count, and processor speed, which collectively indicate a chip performance improvement of around 30%.

Twenty years after his second evaluation, Moore returned to the subject of his law with a final paper.[30] The analysis of complex integrated circuits commercially available in the decade 1965–75 confirmed the initial prediction. Larger die-size and finer line dimension allowed to be achieve about 60% of the projected increase in complexity. As mentioned, the prediction was slightly off because of *memory size*: the planned CCD 256k memories were ultimately not introduced on the market because of problems with information loss. This slowed down the predicted increase after 1975, while decreasing line width stayed on the same exponential trend. The doubling time turned out to be correct, but the number of components per chip was less than predicted. In the meanwhile, the microprocessor and the PC entered the market.

In 1995, the Semiconductor Industry Association predicted that the extrapolation through 2010 of density increase would remain stable, as shown in Figure 9.11. Moore suspected that the trend could not be maintained, mainly for physical and economic reasons: limits to miniaturization below 1.8 micros and the increasing costs of litography production tools growing faster than the revenues were crucial elements to assess the

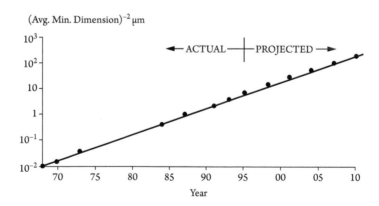

Figure 9.11 Densitiy contribution extrapolation 1995–2010, from Moore (1995)

---

[30] See Moore (1995).

prediction. In its third iteration, Moore returned to stress the economic factors over functionalities:

> I am increasingly of the opinion that the rate of technological progress is going to be controlled from financial realities. We just will not be able to go as fast as we would like because we cannot afford it, in spite of your best technical contributions. When you are looking at new technology, please look at how to make that technology affordable as well as functional.[31]

Hardly any other theoretical prediction in the sciences has had the impact and has been so generally acknowledged as what became known as *Moore's Law*. The prediction has been confirmed in all the successive decades, with a regularity that sometimes has been even suspected to be influencing productions so as to match the law. We have seen how in fact the 'law' was the result of different factors at different times, and how industry used it to direct production. The question whether the law still applies today, or for how long it still will, is an obvious one and it has been posed by many. One factor to be considered is the gain in chip complexity: new materials and methods have to be found in order for the exponential progress to be maintained, and physical or technological limits might determine the end of Moore's era. When engineering fails to achieve in due time the complexity required to meet the exponential curve of the law, a shift in technology will be required, as at that point the law of the market requiring increased functionality and lower costs will be violated and engineering will have to compensate to establish the economical order again. While we do not know yet what such a technological shift might be, manufacturers are already starting delaying production of next generation chips, moving the law's period to 2.5 years.

9.5 Computational Growth

While the evolution of hardware has progressed according to Moore's Law, which since its reformulation in 1995 states that the number of transistors per square inch on integrated circuits doubles roughly every two years, computing ecosystems at large have similar laws, of which Moore's prediction is just one component. In 2017, Denning and Lewis explained how computational growth works:[32]

- at chip level, as formulated by Moore's Law;
- at system level, based on the evolution of interconnection networks, memory systems, caches, I/O systems, cooling systems, languages for expressing parallel computations, and compilers;
- at market level, affected by innovation, adoption, and user communities.

---

[31] (Moore, 1995, p.7).
[32] See Denning and Lewis (2017).

According to this explanation, alongside Moore's Law two parameters should be accounted for in explaining this idea: computational growth at large, and technology adoption.

The speed parameter, essential in measuring system growth including not just chip performance but all other components which might constitute bottlenecks, has been discussed since the 1960s. The first of such laws is known as Amdahl's Law, formulated in 1967: it expresses the theoretical speedup in latency of the execution of a task (velocity of throughput) at fixed workload expected by a system whose resources are improved. The article opens as follows:

> For over a decade prophets have voiced the contention that the organization of a single computer has reached its limits and that truly significant advances can be made only by interconnection of a multiplicity of computers in such a manner as to permit co-operative solution [...] The nature of this overhead (in parallelism) appears to be sequential so that it is unlikely to be amenable to parallel processing techniques. Overhead alone would then place an upper limit on throughput of five to seven times the sequential processing rate, even if the housekeeping were done in a separate processor [...] At any point in time it is difficult to foresee how the previous bottlenecks in a sequential computer will be effectively overcome.[33]

From this initial observation, a law relating speedup and workload is formulated. Let us start with some basic definitions.

Definition 104 (Speedup) *The speed of a program is the time it takes the program to execute. This could be measured in any increment of time. Speedup is defined as the time (T) it takes a program to execute in serial (with 1 processor) divided by the time it takes to execute in parallel (with N processors):*

$$S = \frac{T(1)}{T(N)}$$

Definition 105 (Efficiency) *Efficiency is the speedup divided by the number N of processors used*

$$E = \frac{S}{N}$$

Consider now a task executed before and after some resources are increased. The time T it takes to execute it is built out of

- p: the part that benefits from the improvement of resources, e.g. more parallel processing;
- $1 - p$: the part that does not benefit from the improvement of resources.

---

[33] (Amdahl, 1967, pp.483–4).

The formula for the execution time is then as follows:

$$T = (1-p)T + pT$$

i.e. the time of execution with no improvement plus the time of execution with improvement by parallelization. Let us now assume that resources are improved, with the consequent benefit on the Speedup S:

$$T(S) = (1-p)T + \frac{p}{S}T$$

i.e. the time of the speedup is given by the time of execution with no improvement (which obviously remains the same) plus the time of execution with improvement divided by the speedup (time it takes to execute in serial over time it takes to execute in parallel). The theoretical speedup in latency of the execution time of a task at fixed workload W is then given as follows:

Theorem 19 (Latency Speedup)

$$latency(S) = \frac{TW}{T(S)W} = \frac{T}{T(S)} = \frac{1}{(1-p) + \frac{p}{S}}$$

namely the execution time divided by the time of speedup, both parametrized by the value of the workload; this in turn reduces to the unit of time divided by the time of speedup. As a result of this law, the best speed one could hope for is $S = N$, i.e. where the speedup is proportional to the number N of processors and it would yield a 45 degree curve. That is, if there were ten processors, we would realize a tenfold speedup. When the percentage of the strictly parallel portion of the program is fixed, Amdahl's Law yields a speedup curve which is logarithmic and remains below the line $S = N$. In other words: even when the fraction of work made serially in a program (call it *serial*) is small, the maximum speedup which can be obtained from even an infinite number of parallel processor is only 1/*serial*. Hence the algorithm, and not the number of processors, limits the speedup. As the curve begins to flatten out, efficiency is drastically reduced; see Figure 9.12. What are the effects on programming of Amdahl's Law? First of all, it tells us that substantial parts of code must be executed sequentially in the given compiled sequence; only some of the instructions can be speeded up by parallel execution. Second, only the control-parallel portion of the algorithm can benefit from distinct processors. Finally, it is impossible to achieve much multicore speedup with control parallelism.

This law was reconsidered and challenged in 1988 in an analysis that came to be known as Gustafson's Law.[34] The consideration at the basis of the revision started in an industrial setting, by looking at three problems that were solved with a speedup higher than predicted by Amdahl's Law. Gustafson noted that in Amdahl's analysis there was an implicit assumption that the value of p (the part of work made in parallel) is independent

---

[34] See Gustafson (1988).

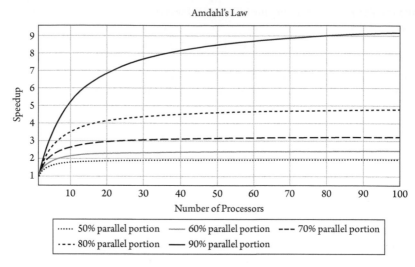

Figure 9.12 Speedup curves according to Amdahl's Law, plotted at https://plot.ly/~benkray/1. embed

of N (number of processors): instead, he considered, in practice the problem size scales with the number of processors and with more processors the problem expands and users can adjust to run the problem in fixed time. Therefore, by assuming runtime and not problem size as a constant, then p scales with the problem and changes linearly with N. Denote now with s (for serial) the speedup of the part of the task that does not benefit from improved system resources; and denote with p (for parallel) the percentage of execution workload of the whole task benefiting from the improvement of the resources. Gustafson's Law states then that the theoretical speedup that can be expected of a system with N processors can be computed as follows:

Theorem 20 (Speedup)

$$S = \frac{(s+p)*N}{s+p} = (s+p)*N = N+(1-N)*s$$

i.e. the speedup is given by the time spent s on the serial work plus the time p spent on the parallel part of the work times the number of processors divided by the overall time. A further identity of Gustafson's Law is that the theoretical speedup in latency $latency(S)$ of the execution of a whole task at fixed execution time T and workload W that can be expected of a system whose resources are improved is as follows:

Theorem 21

$$latency(S) = \frac{TW(s)}{TW} = \frac{W(s)}{W} = \frac{(1-p)W+spW}{(1-p)W+pW} = 1-p+sp$$

What are the effects on programming of this new law? In a data-intensive problem, the data space can be partitioned into many small subsets, each of which can be processed by its own thread: hence, at least for data-intensive applications, adding cores increases the computational work in direct proportion to the number of cores. Rather than paralleliz-ing the algorithm, data-parallel programming parallelizes the data and Gustafson's Law models data-parallel computing, where the speedup scales with the size of the data, not the number of control-parallel paths specified in the algorithm; see Figure 9.13.

The process of parallelizing data rather than algorithms had began in the 1980s. From the 1940s to the 1950s the evolution was from machine level programming to assembly language (no compiling). From the 1960s we moved to high-level languages (compilation) and since the 1990s we have scripting languages. Today, four main paradigms for programming are available:

1. Functional: computations are understood as the evaluation of mathematical func-tions, the behaviour of a functional program can be analysed easily for correctness, there is a heavy stress on avoiding changing states and mutable data in order to eliminate side effects. The functional paradigm has its roots in logic and λ-caclulus in particular; examples of languages of this family are LISP, Racket, Erlang, OCaml, Haskell, F#.

2. Object-oriented: computations are understood as the handling of objects, behaviour, data is combined with procedures to obtain active objects, objects

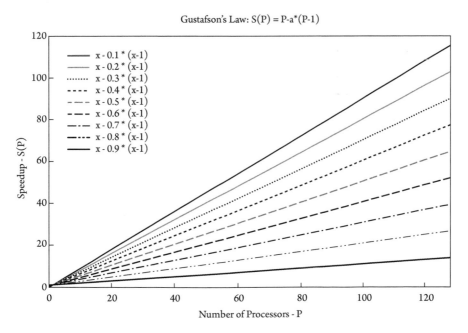

Gustafson's Law: $S(P) = P - a^*(P-1)$

Figure 9.13 Speedup curves according to Gustafson's Law, CC BY-SA 3.0, https://commons.wikimedia.org/wiki/File:Gustafson.png

are grouped by classes with the same behaviours, and classes are hierarchical and satisfy given relations. Examples of languages of this family are Java, Python, C++.

3. Imperative: Machines have re-usable memory that can change state, computations are effected by assignment statements that intervene on the state of the machine, e.g. a statement that assigns value 1 to a memory state x at time t is the value of that memory state at time $t + 1$. Examples of languages of this family are MATLAB, GO (but also Java, Perl, C++).

4. Programming for parallelism, with two main approaches:

   • Implicit parallelism: based on parallel languages and parallelizing compilers; the user does not specify, and thus cannot control, the scheduling of calculations and/or the placement of data but uses automatic parallelization processes; computations use shared memory and there is implicit interaction at compiler or runtime;

   • Explicit parallelism: the programmer is responsible for most of the parallelization effort such as task decomposition, mapping tasks to processors, and the communication structure, using parallel libraries in existing code, requiring major recoding. Common parallel languages used are SISAL and PCN, extensions to or libraries for standard high-level languages, from FORTRAN to C are available as well as message-passing libraries (D, Scala, SALSA).

Of these, parallelism through increased multi-threading is considered often the most relevant and easy to plan for when looking at the future of computing. This is essentially due to the limitations of clock speed imposed by Moore's Law and of novel architectural models not being available yet.[35] But developing for multi-threading is not just a costly process, its complexity also lies in the need for scaling architecture to this aim in order to add flexibility with more cores available. Examples are designs that allow replication of many sets of identical structures on the silicon, and software to turn cores on and off as needed to save power.[36] Both Amdhal and Gustafson's Laws apply to theoretical analyses of the limits of processing with parallel resources. In real applications, the number of such resources is in fact not fixed for any given case, but it varies as a parameter of the efficiency with which multi-threading has been implemented in software. Besides, obstacles to optimization of such processes can be found in the way threads are created and destructed, races and possible deadlocks, load imbalance with threads loaded with heavy computations while others are left with plenty of resources.[37]

This illustrates — in contemporary terms — how the issue of computational growth is always to be considered at system level, taking into account the way in which available technologies, requirements, implementation, and use affect software evolution. This problem is not recent and arose quite early in the history of computing, with its peak in the 1970s and 1980s with the attempt of qualifying computing in terms of software

[35] See Gillespie (2008a).
[36] See Gillespie (2008b).
[37] See (Gillespie, 2008a, p.7).

engineering to deal systematically with this problem.[38] In the 1980s and across the 1990s, a set of laws of software evolution were formulated by Lehman and co-authors to describe a balance between forces driving and slowing progress.[39] Lehman illustrates the problem as follows:

> Programming effectiveness is clearly a significant component of national economic health. Even small percentage improvements in productivity can make significant financial impact. The potential for saving is large. Economic considerations are, however, not necessarily the main cause of widespread concern. As computers play an ever larger role in society and the life of the individual, it becomes more and more critical to be able to create and maintain effective, cost-effective, and timely software. For more than two decades, however, the programming fraternity, and through them the computer-user community, has faced serious problems in achieving this [...]. As the application of microprocessors extends ever deeper into the fabric of society the problems will be compounded unless very basic solutions are found and developed.[40]

Hence, along with economics in line with Moore's analysis, efficiency and effectivenes are identified as main concerns for the correctness of computing systems at large. Lehman distinguishes three types of software systems:

- *S-program*: written according to an exact specification of what that program can do;
- *P-program*: written to implement procedures that completely determine what the program can do (like a program to play chess);
- *E-program*: written to perform some real-world activity, with its behaviour linked to the environment, adapting to varying requirements and circumstances in that environment.

The union class of P and E-programs are denoted as *A-programs*. The first important consequence of this classification affects directly the notion of program correctness. For S-programs, correctness of programs through identity with proofs holds (with the provisos and explanations given in Chapter 7). Real applications and parts of large systems show that this approach is feasible, in the 1980s and even more so today. For E-programs, validity results from human assessment and it cannot avoid taking into account effectiveness. Formal correctness without efficiency becomes irrelevant, while formal incorrectness becomes admissible when compensated for by usability. In several versions, Lehman described the principles guiding the evolution of software for this latter type of system, listed in Figure 9.14.[41]

---

[38] See Chapter 12 for a discussion of computing as an engineering discipline.

[39] See e.g. Lehman (1979, 1980); Lehman and Belady (1985); Lehman (1996); Lehman and Fernández-Ramil (2006).

[40] (Lehman, 1980, p.1060).

[41] See e.g. (Lehman and Fernández-Ramil, 2006, p.31).

Law	Description
Continuing Change	The system must be continually adapted or it becomes progressively less satisfactory
Increasing Complexity	As the system evolves, its complexity increases unless work is done to maintain or reduce it
Self-Regulation	The system evolution processes are self-regulating with the distribution of product and process measures close to normal
Conservation of Organizational Stability	The average effective global activity rate in an evolving system is invariant over the product's lifetime
Conservation of Familiarity	As the system evolves, everyone associated with it must maintain mastery of its content and behaviour to achieve satisfactory evolution. Excessive growth diminishes that mastery; average incremental growth remains invariant as the system evolves
Continuing Growth	The functional content of a system must be continually increased to maintain user satisfaction over its lifetime
Declining Quality	The quality of a system will decline unless rigorously maintained and adapted to operational environment changes
Feedback System	Evolution processes are multi-level, multi-loop, multi-agent feedback systems and must be treated as such to achieve significant improvement over any reasonable base.

Figure 9.14 Laws of software evolution

In general, the laws predict that software grows with linear or even decreasing rate. The topic of the validity of laws of software evolution is still debated today.[42] The longevity of these laws is mainly due to the fact that they have been constantly revised and adapted to

---

[42] See Herraiz et al. (2013) for a review of the literature.

account for the evolution of computing systems. Studies aimed at verifying the validity of the laws have been made, for example, on open-source software.[43] Some software projects present a majority of cases which invalidate the laws entirely, with software showing superlinear growth. The difference is often due to what definition of evolving software is taken into account.

Besides the divergence in their empirical validation, these laws illustrate how validity and correctness for software respond to criteria of change and update (the law of continuous change seems in general validated by all studies). Studies on standard software systems have invalidated the view that changes induce degradation, in turn increasing complexity: the overall complexity of computational architectures and the extensive introduction of new programming paradigms (e.g. parallelism) certainly contribute to strengthening the levels of complexity. Hence it is crucial to identify which systems qualify for these laws: one factor that seems to affect negatively the validity of the laws, and in particular the prediction of linear or negative growth on complexity, is the distribution of processes, architectures, and contributors. Exploiting explicit parallelism for reducing power consumption means increasing both cores and performance, or increasing cores and decreasing frequency: but programming for multi-core has several bottlenecks, most notably memory, with floating point operations per second coming for free, but bandwith being expensive. To deal with the issue of memory bottlenecks, the architecture has changed in a hierarchical form.[44] The memory hierarchy requires a principle of locality for data, both temporal and spatial, formulated by two simple principles:

- a memory location that was referenced in the past is likely to be referenced again;
- a memory location next to one that was referenced in the past is likely to be referenced in the near future.

The design of parallel algorithms requires keeping track of such locality in the memory and which parts of the caches are accessible (in terms of rights).

This analysis illustrates—in an engineering context—the essential aspect of controlling processes at local and global level. Computational growth and increasing complexity of the software and hardware architectures have further side effects, namely the increasing difficulty of preserving functionality. This results often in computational errors or miscomputation. These aspects, including how computing practice aims at reducing the risk of computational errors at several levels of abstraction, and how such errors manifest themselves when not avoided, will be analysed in the next chapters.

---

[43] For the Linux kernel for example, mixed results have been obtained: this highly complex system is considered compatible with the laws, but showing some laws contradicted (increasing complexity, conservation of familiarity, and declining quality) and some only partially validated (self-regulation, conservation of familiarity for some releases, feedback system); see Israeli (2010) and (Herraiz et al., 2013, p.13).

[44] See e.g. (Culler et al., 1998, ch.3.2).

Exercises

Exercise 92 *Offer a general outline of the working of a transistor.*

Exercise 93 *Offer a general outline of the working of magnetic core and mention its properties.*

Exercise 94 *Provide an overview of the evolution of microprocessors according to the relevant eras.*

Exercise 95 *What does Moore's prediction establish? Which elements were of relevance in its first formulation?*

Exercise 96 *How was Moore's prediction modified in its successive formulations and which aspects were considered relevant?*

Exercise 97 *Which aspects affect the evolution of a computational ecosystem at large?*

Exercise 98 *Explain the difference between Amdahl's and Gustafson's Laws.*

Exercise 99 *Explain the distinction among programs according to Lehman.*

10 Properties of Implemented Computations

Summary

In Chapter 9, we considered the historical evolution of computing machines from the second to the fourth generation, highlighting several principles and laws of computational growth. In this chapter, we consider how data formulation, hardware architecture, and errors related to both essentially characterize computations at several levels of abstraction from algorithms to implemented programs.

10.1 Physical Computing

The structural definition of a physical computing device based on the von Neumann architecture rests on the interpretation of information processing by algorithmic steps with I/O operations. This reading has two important caveats for the correctness of an *implemented* program: first, the inputs and resources from which the computation depends need to be accessible at runtime, and of the type and form required; second, the physical architecture on which the computation is performed should not impede its correct execution through limitations or malfunctioning. Accordingly, the purely formal problem of establishing what is a valid computation expressed in terms of program correctness becomes the engineering problem of establishing what can be computed correctly by a physical device. This, in turn, leads to the issue of what can be computed efficiently. The theoretical and practical importance of this problem establishes the convergence between mathematical and physical computing. It is essential to determine whether a function which cannot be implemented (or not efficiently so) can still be considered computable. This problem applies by extension to every new proposed model of computation. Our main concern in this chapter is to establish properties of the engineering foundation of computing which play a crucial role for the associated notions

On the Foundations of Computing. Giuseppe Primiero, Oxford University Press (2020). © Giuseppe Primiero.
DOI: 10.1093/oso/9780198835646.001.0001

of correctness and validity. To this aim, establishing which physical systems qualify as computational and to identify which processes can be executed by such systems become important philosophical and technical problems.[1]

An early general account of physical computation due to Putnam defines anything that is accurately expressed by a computational description C (e.g. the instruction table of a Turing Machine) as a computing system implementing C (i.e. the machine implementing the program expressed by that table).[2] This view has been widely discussed and it has become known in the literature as the simple mapping account:[3]

Definition 106 (Simple Mapping Account) *A physical system S performs computation C if*

1. *there is a mapping from (possibly a subset of, or equivalence class of) the states ascribed to S by a physical description to the states defined by a computational description C, such that*

2. *the state transitions between the physical states mirror the state transitions between the computational states.*

In clause 1. the problem is to restrict the states of S to significant ones and to group them according to an equivalence relation (similarly to what was discussed for the notion of algorithm in Chapter 6.6). Clause 2. requires that for any computational state transition $s_1 \rightarrow s_2$ specified by the computational description C, if the system is in the physical state that maps onto s_1, it then goes into the physical state that maps onto s_2.

Various approaches offer significant restrictions on the states of the system considered for the computational description:

1. *Causal interpretation*: a causal relation is required between the physical states, so that for any computational state transition of the form $s_1 \rightarrow s_2$ (specified by the computational description), the physical state that maps onto s_1 causes the system to go into the state that maps onto s_2.[4]

2. *Syntactic interpretation*: If a syntactic restriction is required on physical states, then only states that qualify as syntactic are relevant to the computational description, i.e. if a state lacks syntactic structure it is not computational; then a computation consists of the manipulation of language-like expressions sensitive to syntax.

3. *Semantic interpretation*: If a semantic qualification is required on the states of the physical system, then only states that qualify as representations are relevant to the computational description.[5]

---

[1] In this section we do not aim at offering a full overview of the philosophical positions on physical computing. See instead Piccinini (2017).

[2] See Putnam (1975).

[3] See e.g. Godfrey-Smith (2009).

[4] See e.g. Chrisley (1995); Chalmers (1995); Scheutz (1999).

[5] See e.g. Pylyshyn (1984); Churchland and Sejnowski (1992); Shagris (2006).

4. *Linguistic interpretation*: A representation can be intended as linguistic manipulation and what counts as a computable structure requires a language-like
combinatorial structure in a way that is sensitive to their syntactic structure and
preserves their semantic properties (e.g. compositional semantics).[6]

5. *Mechanistic interpretation*: A minimalist account of physical computation, not
requiring to endorse either syntax or semantics, can be given in terms of functional
mechanisms: a computation in the generic sense is the processing of elements of a
computational system according to rules that are sensitive to specific components
of those elements. The processing is performed by a functional mechanism,
that is, a mechanism whose components are functionally organized to perform
the computation. Thus, if the mechanism malfunctions, a miscomputation
occurs.[7]

A possible limitation with the mechanistic definition of computation is highlighted
precisely by the notion of miscomputation. While a mechanism can be defined more
or less abstractly, renouncing entirely to a semantic interpretation (and in particular
renouncing to an intentional reading of the function implemented by the computational
system) makes it impossible to distinguish between the different reasons according to
which a computational system may malfunction: a computational error as a side effect,
an error in terms of program non-termination, or an error induced by the system not
returning the intended output.[8] The notion of specification requires therefore necessarily
an act of interpretation, which can be nonetheless integrated in a minimalist mechanistic
account of physical computation.[9] This formulation of the mechanistic account is viable
only as long as it is complemented with an interpretation of correctness of the processing
rules with respect to an intended specification and with an understanding of efficiency.
An analysis of the notion of correctness and validity in physical computation systems
preserving both a syntactic qualification of its states and a semantic interpretation of its
intended specification seems therefore required.

Such interpretation can be offered by remaining closer to the nature of physical computing in terms of information processing formulated at the beginning of this section.
Existing interpretations along these lines have been criticized because the definition
of a computational process seems to become too dependent on the assumed notion
of information: physical, probabilistic, syntactic, semantic notions of information, each
seems to return a different definition of computational process. This limitation results

---

[6] See e.g. Fodor (1975). Note that in Piccinini (2017) this thesis is rejected. We maintain that the set
of computable function schemata being translatable to a context free grammar is a sufficient indication that
computation presupposes a language-like combinatorial structure.

[7] See e.g. Piccinini (2015); Milkowski (2013); Fresco (2010, 2014).

[8] We will consider this distinction more thoroughly in Chapter 11, defining miscomputations in terms of
misfunctions and disfunctions. For the purposes of the present chapter, it is interesting to note that the definition
of a *class* of well-functioning mechanisms for an intended specification is required in order to be able to address
the notion of efficient computation.

[9] For example, for a mechanistic account of computing systems where malfunction is induced through
malicious software aimed at modifying the system's intended specification, see Primiero et al. (2018b).

from a partial account of information processing in computing. Instead, a comprehensive interpretation of physical computing as information processing requires preserving all the distinct levels of abstraction presented by a computational system. The levels are summarized in Figure 10.1. Each level can be defined in terms of the domain of object it refers to, presented in Figure 10.2. Each layer in this ontological domain is associated with the one above it: an electrical charge is controlled by machine code, which is denoted by a programming language construct, which implements an algorithm, which satisfies an intention. The explanation of the ontology requires an appropriate epistemological structure, presented in Figure 10.3. Each epistemological construct has a relation with the underlying one: a problem is reflected by a task, which is interpreted by an instruction, satisfied by an operation and executing an action.

This structured presentation of the ontology and epistemology of physical computing systems allows to be expressed for each level a relevant notion of information.

Level of Abstraction	Description
FSL	Intention of Requirements Specification
DSL	States Description of the Specification
ADL	Rules Description of the Specification
AIL	High-/Low-level Translation of the Algorithm
AEL	Execution in Hardware

Figure 10.1 Levels of abstraction in computational systems, Fresco and Primiero (2013)

Ontological Domain
Intention
is satisfied by
Algorithm
is implemented in
Programming Language
denotes
Machine Code
controls
Electrical Charge

Figure 10.2 The ontological domain of physical computing systems

Epistemological Structure
Problem
is expressed in
Task
is interpreted in
Instruction
is satisfied by
Operation
is executed in terms of **Action**

Figure 10.3 The epistemological structure of physical computing systems

In particular, for each epistemological construct, an associated form of information processing can be formulated:[10]

- at Functional Specification Level, information processing is intended as problem-solving: the corresponding notion of information is abstract, correctness- and truth-determining information in terms of intentions satisfied by algorithms;

- at Design Specification Level, information processing is intended as task formulation: the corresponding notion of information is abstract, correctness-determining information in terms of algorithms implemented by instructions;

- at Algorithm Design Level, information processing is intended as instruction formulation: the corresponding notion of information is well-formed meaningful data in terms of the relevant high-level language constructs denoting operations;

- at Algorithm Implementation Level, information processing is intended as operation execution: the corresponding notion of information is well-formed performative data in terms of machine code controls actions;

- at Algorithm Execution Level, information processing is intended as executed action: the corresponding notion of information is syntactic data in terms of electric charges, logically structured and executed.

Each level constitutes the ontological domain for the level above; and each level expresses the epistemological structure for the level below. For example, the domain of electric charges on circuits is purely syntactic data structured by logical rules; this is the domain of objects controlled by machine code. In turn, the operation performed at low-level language executes the action expressed by the logical structure of gates and other components.

As each level of abstraction implements a different notion of information, an informational definition of physical computing has to account for all of them:

Definition 107 (Informational Account of Physical Computation) *A physical computation C is the algorithmic processing of information sensitive to the level of abstraction at which a physical system S is formulated. If processing fails, a miscomputation occurs at a given level of abstraction.*

Note that miscomputation is here no longer a generic malfunctioning but is qualified at the appropriate level of abstraction.

The problem of defining physical computing in turn puts under scrutiny the general validity of the Church-Turing Thesis. Recall from Chapters 4 and 5 that the thesis establishes that all effectively computable functions are computable by some Turing Machine. For the analysis of physical computational processes, two versions of the thesis have been offered in the literature. The first one, is the strongest version:[11]

---

[10] For this interpretation, see Primiero (2016).
[11] See Deutsch (1985).

Definition 108 (Bold Physical Church-Turing Thesis) *Every finitely realizable physical system can be perfectly simulated by a universal model of a computing machine operating by finite means.*

According to this first, stronger formulation, for any finite system it is possible to provide a computational description which reduces to a Turing Machine's instruction table. A first critique of this account relies on the strength of the simulation relation it subsumes: a strong relation leads to a form of pancomputationalism; a weak relation risks allowing simulations which are not faithful. In our terms, a further critique can be formulated: no connection is made explicit in the simulation relation between the actual levels of information processing for the physical system and the abstract level of the Turing Machine model.

A weaker formulation of the thesis is as follows:[12]

Definition 109 (Modest Physical Church-Turing Thesis) *Any function that is implemented in a physical system can be simulated by some Turing machine.*

This second formulation relaxes the criterion of perfect simulation of the bold version and indirectly refers to processes *usable* by an observer to generate values of a function within a physical system, where usability can be specified by several constraints.[13] One problem is therefore to define such notion of usability. Moreover, an additional weakness of this formulation of the thesis is that it still does not qualify the relation between abstract algorithm and its implementation.

In the following, we shall reconsider modern physical computing systems in terms of their essential properties. Our aim is, based on practical and theoretical results, to provide analytic criteria for these notions and use them to return to the problem of computational correctness and the appropriate corresponding version of the physical Church-Turing Thesis.

10.2 Functionality

A first property which we naturally identify with physical computing is functionality. At high level, this is usually assumed to be a software property, and it could be even reduced to correctness: the system is functional when it does what it is supposed to do. We have previously considered to some extent software correctness as a formal property, and we will return to the issue later in Section 10.3 of this chapter in terms of (hardware and software) systems working in the presence of errors and in Chapter 11, where correctness will be approached from the point of view of implementations. For the time being, we

---

[12] Works endorsing the modest version of the physical Church-Turing Thesis are for example Smith (2006); Beggs and Tucker (2007).

[13] The mechanistic account proposed in (Piccinini, 2015, chs.5–6) and Piccinini (2017) refers to the following constraints: I/0 readability without error; fixed, process-independent rules; repeatability of the process; settability; physical constructibility; reliability.

consider functionality at a lower level of abstraction: when we explicitly consider physical computation, functionality refers to data validity and hardware control.

Let us start by providing a basic definition of data validation:

Definition 110 (Data Validation) *Data validation denotes the process aimed at showing systematically which data a given program can and cannot process correctly.*

This aim is reached by a systematic methodological approach, which consists in the process of data being checked, accepted, or rejected against an established set of criteria.[14] The several existing techniques for data validation are categorized into two main methods:

- manual inspection: analysts and technicians inspect results for values that are higher or lower than expected, or appear to be outside control limits (outliers);
- computerized techniques.

Data errors are usually defined in terms of inconsistencies, incomplete data, and the discovery of outliers. To detect (automatically) faulty data, a number of standardized tests can be performed; see Figure 10.4. Once faulty data has been identified, data correction methods are put in place. Examples of such methods are:[15]

1. Interpolation: insert data where missing (typically the last available);
2. Smoothing: change non-representative data with averages;
3. Data mining: extract values from data in case of wrong values;
4. Reconciliation: use equations to adjust data.

By the standards of modern personal computing, data validation is a process that the operating system delegates to the relevant program by the use of routines in the appropriate language. Such routines allow the computer applications to test data to ensure their accuracy, completeness, and conformance to industry or proprietary standards. This can happen through the implementation of strict data typing or data structuring methods, or by the implementation of simple informative error-escaping protocols. Such methods are nowadays defined by data quality rules, according to one of the following categories:[16]

- *Attribute Domain Constraint*: used to restrict the allowed values of individual data attributes;
- *Relational Integrity Rules*: used to enforce identity and referential integrity of the data, e.g. by record linkage;
- *Rules for Historical Data*: used to ensure timeline constraints and value patterns for time-dependent value stacks;

---

[14] See e.g. (Ratliff, 2003, p.19).
[15] See (Sun et al., 2011, pp.11–14).
[16] See Maydanchik (2007).

Test	Explanation
Physical range check	physically possible values
Realistic range detection	realistic values usually observed
Gap detection	establish when data are missing
Constant value detection	period the values are constant
Signal's gradient	sudden or erratic change of value
Tolerance band	admitted outliers
Material redundancy	comparison of redundant values
	for unusual values or gaps
Physical or mathematical models	modelling for expected values
Statistical check	control of extreme values
Drift detection	change control
Spatial consistency	location control
Analytical redundancy	definitional identity
Gross error	discrepancy of observed
	from expected statistical value
Data mining	extracting patterns for diagnosis

Figure 10.4 Standard test for faulty data, (Sun et al., 2011, pp.6–10)

- *Rules for State-Dependent Objects*: used to constrain the life-cycle of objects described by their state-transition models;
- *General Dependency Rule*: used to identify complex relationships between data.

For mainframe computers, data validation is a process that the OS has to manage. As a modern example, consider the IBM z/OS 64-bit operating system, initially released in 2001 and currently operative for high-volume operations with high data sensitivity and security levels. A Java program called the *IBM Content Manager* is deputed to validation of data through inconsistency checking of several kinds (missing reference, missing referenced, size mismatch, collection mismatch, date mismatch, name discrepancy).

With data quality assessment one thus refers to the process of evaluating data to identify errors and understand their implications. Despite the large number of practical approaches and the development of tools to help with data quality modelling, identification, and error resolution, a major open issue concerns the foundational problem of defining a reference set of data quality dimensions and metrics. Among others, fitness for purpose is considered central: it is analysed in terms of its subjective vs. objective value and its domain dependence.[17] A standard set of data quality dimensions includes:[18]

- *Accuracy*: the closeness between an available value v and an ideal value v' considered as the correct representation of the real-life phenomenon intended by v;

[17] See (Batini and Scannapieco, 2006, pp.221–2) and Illari (2014).
[18] The earliest categorization of information quality dimensions appeared in Wang (1998); for a later and more extensive categorization, see e.g. Lee et al. (2002). The following definitions are extracted from Batini and Scannapieco (2006).

- *Completeness*: the extent to which data are of sufficient breadth, depth, and scope for the task at hand;[19]
- *Timeliness*: how current data are for the task at hand;
- *Consistency*: whether semantic rules defined over (a set of) data items are violated, where items can be tuples of relational tables or records in a file;
- *Integrity*: protection of the information from unauthorized, unanticipated, or unintentional modification, to prevent it from being compromised by corruption or falsification;
- *Accessibility*: the ability of the user to access the data from his or her own culture, physical status/functions, and technologies available.

This first aspect of functionality related to data validity and correctness can thus be summarized by saying that the system must be effective (i.e. work properly) when the input data is correct and of sufficient quality. But guaranteeing the quality of input data is only a first step towards functionality.

A second essential step is testing that hardware is working correctly. Again, we start by providing a basic definition:

Definition 111 (Hardware Testing) *Hardware testing denotes the systematic check of the physical well-functioning of computer hardware parts.*

Hardware testing ensures that basic units of a computing system respect design objectives and functionalities as by specification, consisting essentially of:

i) low-level testing, e.g. in the form of stress-testing of unit components to determine their stability; it involves testing beyond normal operational capacity to determine breaking points or safe usage limits, to confirm intended specifications are being met, and to determine modes of failure;

ii) high-level (functional), possibly automated testing of very large-scale integration circuits; the ratio between good working and produced ICs is always below 100%, with probability of circuits failing to meet their specification at any point in the manufacturing process increasing in proportion to size and complexity; here testing is required for the functional design, the wafer probe, the final functionality.[20]

After testing is performed at manufacturing level, system usability is considered to include usefulness, efficiency, effectiveness, satisfiability, learnability, and accessibility.[21] Metrics used to quantify system behaviour are:[22]

---

[19] See e.g. Wang and Strong (1996).

[20] See e.g. Grochowski et al. (1997).

[21] See e.g. Rubin and Chisnell (2008).

[22] For mainframe systems, the first three are the main standard criteria defined by IBM, called RAS model, see https://www.ibm.com/support/knowledgecenter/zosbasics/com.ibm.zos.zmainframe/zconc_RAS.htm.

1. *Reliability*: the probability that the output will be correct up to a certain time t, as a result of the ability to make quick updates on detected problems;

2. *Availability*: the amount of time t the system is operating, as a percentage of total time it should be operating, also called *mean time between failure* (MTBF);

3. *Serviceability*: the speed at which the system can be repaired in case of fault, also called *mean time to repair* (MTTR), implying that units of hardware and software replacement are well defined;

4. *Safety*: the probability that the system either will be operating correctly or will fail in a 'safe' manner at given time t;

5. *Performability*: the probability that the system is operating correctly at full throughput or operating at a reduced throughput greater than or equal to a given value at time t;

6. *Testability*: how easy it is to test the system.

If a system is said to have 99.999% of availability, it means it must not be unavailable for more than five minutes over the course of a year: this is called *high availability*. The same law of *five nines* is what defines *high reliability*, with every single component producing a high impact on overall reliability: a single component out of 10 whose reliability drops to 95% makes the overall reliability value of the system drop to 94.99%. Fault tolerance is the ability of the system to continue correct operation after the occurrence of hardware or software failures or operator errors. High availability and fault tolerance are obtained by redundant hardware components in critical paths, enhanced storage protection, a controlled maintenance process. From the software point of view, the system must be designed for unlimited availability. These techniques allow to minimize the risk of having a single point of failure for a whole computer system. A balance between reliability and availability is usually sought, as redundancy does not scale very well in large systems: hence, higher costs for more reliable components, which can be estimated based on testing and expected failure rates, usually pay off with large systems. In this context, system architects take into account the likelihood of unit failure, the impact of failure on the system and the cost of recovery versus the cost of fail-safe isolation.[23] Reliability is often favoured over availability in safety-critical systems, like aircrafts and power plants.

We summarize the analysis above in the formulation of a functionality principle for physical computing:

Principle 9 (Functionality) *A functional physical computing system is able to satisfy the aims and objectives for which it has been designed, assuming validity of the data processed and a level of hardware reliability and availability.*

The central principles of limited correctness are introduced in Section 10.3 under the concept of *usability*.

---

[23] See (Siewert, 2005, p.4).

10.3 Usability

Reliability and availability in physical computing point to another crucial aspect: the issue of errors and their handling in software and hardware. This is a problem of enormous relevance for both theoretical and practical reasons, with consequences for security, fault prediction, formal verification, and repair strategies. The second principle of physical computing is usability, and it refers to the criterion of preserving a certain level of functionality also in the presence of possible errors.

From an applications-oriented viewpoint, identifying and resolving faulty states is a major concern especially in the context of systems with distributed resources, contents, devices, services. From an engineering viewpoint, error detection, fault tolerance and handling are well-studied notions since the 1970s.[24] Typical cases of faulty operations in distributed systems are related to resource sharing (as in communication), resource typing and accessibility (as in database theory and many physical processes). The structure of distributed systems, with concurrent processes without global clock, is responsible for generating such independent failures. On the other hand, designing rules and meta-theoretical properties in formal systems to study validity and satisfiability conditions is an essential theoretical enterprise. Errors in formal contexts are avoided rather than resolved, and this can be recognized in real-world applications as well. The standard approach to fault tolerance strategies in real distributed systems consists in detecting the error, categorizing it as solvable or not, and identifying the affected service and its criticality. Finally, assessing possible resolutions strategies or service restart. This type of approach towards failure handling is standard in both the functional and in the object-oriented programming paradigms. In Java, for example, exception handling relies on the distinction between checked exceptions (a user error or a problem that cannot be foreseen by the programmer, examples are `ArrayIndexOutOfBound Exception` and `ClassCastException`), runtime exceptions (an exception that could have been avoided by the programmer, examples are `FileNotFoundException` and `ParseException`) and errors (problems beyond the control of the user or the programmer, examples are `OutOfMemoryError` and `StackOverflowError`). In Java, both runtime exceptions and errors are unchecked exceptions and by definition they are ignored while checked exceptions are those from which a program can recover. A programmer is expected to check exceptions by using the try-catch block or throw it back to the caller. For errors the programmer is not required to do anything and the program terminates. Python offers another example of exception handling of the same kind, maybe even weaker. Exceptions, which include Errors as a kind, are comparable to Java's Throwable Class of exceptions, but they are not all checked and the programmer is not forced to do all the checking work at compile time. The distinction between solvable and non-solvable exceptions can be found also in functional programming, for example with the distinction between the `Terminate` and `Resume` exceptions, where only

---

[24] See Goodenough (1975); Cristian (1982); Rennels (1984); Cristian (1985, 1991).

the second applies to 'curable' exceptions, implemented in the `Throwable` data class in e.g. Scala.[25]

For hardware, a deviation from standard physical behaviour in a computational system is called a *failure*, and it can be temporary (transient faults that appear at a particular time and disappear after some period, or intermittent faults occurring from time to time) or permanent (faults that remain in the system until repaired). A failure can be attributed to incorrect functions of one or more physical components, incorrect hardware or software design, or incorrect human interaction. Incorrect data or signal output is called an *error*, and it can be the result of a failure. Errors without effects on system's functioning are possible and will be considered in Chapter 11. *Fault models* are designed to represent the effect of a failure by means of the change produced in the system signals.[26] Techniques like modelling, verification, and testing aim at eliminating any possibility of faults. For many applications, especially non-critical ones, errors on output are admissible. In these cases, *fault avoidance* is the technique aiming at reliability based on increasing mean time to failure in terms of: robust (formal) design, (formal) design validation verification, reliability verification and production testing.

Fault-tolerance in computing systems aims at limiting the effects of malfunctioning on the system's output, by providing responses that make the system work to some standard even in the presence of errors. The introduction of fault tolerance is based on the unavoidability of errors for large-scale systems: on the one hand, formal verification relies often on an abstraction on system states that cannot pre-empt all possible error conditions; on the other hand, testing is never exhaustive, with realistic conditions hard or impossible to obtain, accuracy in simulation difficult to reach and errors introduced at the requirements stage possibly manifesting only when the system is operational. Where errors on output are not admissible, like in real-time control systems, fault tolerance is expressed as error-masking and it expresses the total number of components that can fail (simultaneously or not) without the system presenting error on output.

While full fault tolerance admits errors in the preservation of full functionality, *graceful degradation* refers to continuing operativity in the presence of errors, accepting a partial degradation of functionality or performance during (manual or automatic) recovery or repair through the removal of failed components.

Finally, *fail safe* refers to the situation in which the system maintains its integrity while accepting a temporary halt in its operation. Fault secure techniques guarantee output correctness unless an error indication is provided; when errors are detected they need to be corrected, hence making it adequate for situations that allow recovery and retry processes (e.g. banking, telephony, networking, transaction processing, etc.).

The most common among the various automatic failure management techniques is identification of faulty components through built-in testing, and disconnection upon reconaissance. Self-repairing techniques are available in systems whose accessibility is too costly or impossible (e.g. for satellites), consisting mainly in:

---

[25] See e.g. Govindarajan (1993).
[26] For this categorization of faults, see (McClusky and Mitra, 2004, secs2.1–3).

- *dynamic redundancy*: redundancy supplied inside a component which indicates that the output is in error and recovery provided by stand-by spares;
- *hybrid redundancy*: combination of spares with a scheme of three modules whose outputs are passed through a voting network before being transmitted as outputs called triple modular redundancy (TMR): the outputs or votes should be identical, consensus is taken to be the correct output;
- *static redundancy*: TMR alone.

Static redundancy software is problematic in terms of how often voting should be performed and to what extent consensus should be reached. Data integrity can be guaranteed by duplication, where the same logic function is implemented twice and correctness is assumed by agreement on their outputs. When disagreement is present, fail-safe mechanisms allow the system to revert to a previous state.

Failure recovery is the process that involves restoring an erroneous state to an error-free state when failure is assumed to be no longer temporary. *Forward Error Recovery* relies on continued functioning from an erroneous state, by making selective corrections to the system state through making the controlled environment safe. *Backward Error Recovery* aims at restoring the system to a previous safe state and executing an alternative section of the program. Besides the more drastic rebooting, rollback recovery is the most common technique. It can be performed with two approaches:

1. operation-based recovery: all modifications to the system have been recorded so that a previous state of the process can be restored by reversing all the changes;
2. state-based recovery: the complete state of a process is saved at various checkpoints.

Checkpoints can be created: i) by state-based or operation-based, with all checkpoints saved in stable storage; ii) by global checkpointing, with actions coordinated such that each process saves only its most recent checkpoints, and the set of checkpoints in the system is guaranteed to be consistent. The requirement on the consistent recovery state is that every message that has been received is also shown to have been sent in the state of the sender (i.e. strongly consistent set of checkpoints). Recovery assumes that processes have regularly been checkpointing their state, so that the most recent consistent global checkpoint is chosen. Note that this consistency requirement assumes reliable communication, while an alternative is represented by message logging.

Fault tolerance expresses, in other words, the criterion of usability in the presence of possible, in principle unavoidable but irrelevant errors to the aim of (overall) functionality.

Principle 10 (Usability) *A usable physical computing system guarantees services delivery and uses resources within admissible limits and in the presence of possibly unavoidable errors, but which are irrelevant to the overall functionality, or from which the system is able to recover.*

In the informational account of physical computation (cf. Definition 107), usability also includes the ability to set the process according to the appropriate level of abstraction:

as a problem-solving process, task, instruction set, low-level operations, syntactic actions; at each such level, the appropriate set of variables (the so-called gradient of abstraction) can be determined and interpreted to evaluate usability. To complete the picture of physical computation, we need to address a final aspect along with functionality and usability, namely efficiency in terms of execution-time and energy.

10.4 Efficiency

Under their formal expression, computing processes not connected to a real physical environment (like the Turing Machine model) do not consider efficiency as relevant to reliability and usability. The abstract machine model expresses computation optimization in terms of time (how long does the algorithm take to terminate) and space (how much memory is needed by the algorithm). This translates physically in terms of RAM, data alignment and granularity, word sizes and clocks.[27] Efficiency is traditionally a parameter of computational resources used by the algorithm, with maximum efficiency correlated to minimum resource usage. In this setting, usually average execution time is considered and the O-notation ('big-O') expresses the upper-limit complexity of an algorithm as a function of the size n of the input replacing real-time units; see Figure 10.5. For space complexity, the amount of memory needed is usually defined in terms of:

- amount needed by the encoded algorithm;
- amount needed by the input;
- amount needed by the ouptut;
- amount needed by the computation process.

Against this background, for computing systems that are integrated in a physical environment optimization stresses functionality and reliability more than efficiency: this has led to little effort made in the formulation of precise timing requirements in the evolution of embedded software. Timing has been sacrificed for scalability, context-dependency, and Turing-completeness. Termination is essential, but not decidable and this model does not interact strongly with timing, which is required to prove termination of subprocedures.

$O(1)$	constant
$O(\log n)$	logarithmic
$O(n)$	linear
$O(n \log n)$	quasilinear
$O(n^2)$	quadratic
$O(c^n), c > 1$	exponential

Figure 10.5 The main classes in the O-notation

---

[27] See Lee (2005).

Physical critical systems make essential use of concurrency and temporality. Timing in software is often abstracted in terms of ordering and threads, the latter being essential to formulate virtual parallelism. This technique, based on semaphores and mutual exclusion locks, is risky in view of possible failures and non-deterministic behaviours due to races, which are obviously not acceptable in critical systems. In this setting, standard testing and static analysis techniques are often not sufficient to guarantee high reliability. Recall that the problem of partial correctness is expressed in terms of termination and valid output. Termination in a given time interval has been formulated either as upper bounds for safe computation optimizing on worst-case execution time (WCET), or as execution time independent of input values, with the latter often too strict a criterion. Predicting timing happens either by programmable digital signal processors or by software interacting with device-specific hardware.[28]

The other important factor alongside computation optimization is power or energy efficiency. Especially in industrial contexts, both the physical infrastructure and the computing technology represent crucial problems as high-performing computing represents an enormous consumer of energy. US datacenter energy consumption in 2006 was estimated at ~ 61 billion kWh, for a total cost of $4.5 billion, more than twice as high as in 2000. In 2012, ICT usage consumed 4.7% of electricity worldwide, ~ 920 terawatt-hour, i.e. 10^{12} watt-hours (overall worldwide electricity consumption is 15% of the total produced). In 2008, the projected annual growth in worldwide electricity use for the period 2002–20 was estimated up to an annual 12% for datacenters only. In 2010, stated electricity consumption by datacenters had risen by 56% in five years.[29] A law of efficiency growth has been postulated:[30]

Definition 112 (Khazzoom-Brookes Postulate) *Energy efficiency at the micro-level leads to higher energy consumption at the macro-level.*

Despite being formulated first for household appliances, this postulate is assumed to faithfully describe the evolution of high-performance computing: energy is becoming cheaper, energy efficiency leads to economic growth, and increased consumption in one technology produces bottlenecks that lead to increased use of other technologies; the latter leads back to increasing energy consumption. Data reported above seems to confirm this trend. Performing computations are thus sought against energy costs and with increasing CO2 impact concerns. The trade-off with the minimum energy consumption has already been studied also in terms of data transmission methods over distributed networks.[31] Other energy-efficient methods involve program and network designs.

We conclude our characterization of physical computing system with a principles of efficiency:

---

[28] See Marwedel and Engel (2012) for an overview of techniques and hardware.
[29] For these data, see Consortium (2013); Van Heddeghem et al. (2014); Agency (2012); Pickavet (2008); Gelenbe and Caseau (2015); Koomey (2011).
[30] See Khazzoom (1980).
[31] See Gelenbe and Lent (2013); Gelenbe and Morfopoulou (2011); Sakellari et al. (2013).

Principle 11 (Efficiency) *An efficient physical computing system delivers functionalities at least as fast as required, parametrized by computational resources.*

A usable, efficient system is preferable to a fully correct but inefficient system.

10.5 Limits of the Church-Turing Thesis

In the analysis above, we have highlighted three main principles which can be considered at the basis of physical computation: functionality, usability, efficiency. This reading of physical computing in the context of an informational view distinguishing levels of abstraction allows one to reformulate the debate on the bold vs. modest versions of the physical Church-Turing Thesis. The advantage of the present framework is to allow for a precise distinction of the levels of abstraction at which algorithmic function and physical system are located. With the distinction between information processing as tasks (algorithms), as instructions (high-level programming language), as operations (low-level programming language), and as actions (data execution) and our previous analysis on the different meanings of algorithms, it seems clear that the abstract computational model of the Turing Machine can only be used to simulate computational processes at the higher level of abstraction, as it accounts only for functionality, but it does not express either usability or efficiency. In this light, when considered at the highest level of abstraction, a formulation of the algorithmic Church-Turing Thesis is possible:

Definition 113 (Algorithmic Church-Turing Thesis) *Any function expressing at the algorithmic level a task for a computational system can be fully simulated by some Turing Machine.*

The two relata of the simulation occur at the same level of abstraction and therefore there is no qualification required on their relation. At this level there is no implementation in a physical system.

When the notion of computational system is intended in its physical realization, the notion of simulation has necessarily a weaker meaning. We propose here that the relation of equivalence between algorithms, instructions, and operations can be accounted for in terms of the linguistic equivalence of the respective models:

Definition 114 (Linguistic Church-Turing Thesis) *Any function expressing at the algorithmic level a task for a physical computational system which can be functionally implemented at the instruction level (e.g. by a high-level programming language) and efficiently implemented at the operation level (e.g. by a low-level programming language for a given architecture) can be linguistically simulated by a Turing Machine (in an appropriate instruction table minimal for the computation at hand).*

The simulation at hand expresses the correspondence between the algorithm, the instructions and procedures of a physical computational system with an appropriate program for a Turing Machine. At this stage, functionality is accounted for in terms of correct output and efficiency in terms of minimal alphabet and instruction table. But there is no full account of efficiency beyond computation optimization, i.e. there is no account of power or energy efficiency as outlined above. Moreover, the problem of usability as expressed by fault-tolerance procedures is entirely missing.

The Turing Machine model abstracts entirely from aspects of computation within admissible limits and preservation of service delivery (even at the cost of full correctness), which we have stressed for physical computing. In order to take these aspects into account, it is essential to offer an analysis of miscomputations highlighting the different levels of abstractions at which errors can occur. Such layered analysis has to consider, first of all, the gap between the formal specification and its implementation in terms of physical realizability. These aspects are the aim of Chapter 11.

Exercises

Exercise 100 *Explain the simple mapping account and its constraints.*

Exercise 101 *List the levels of abstraction for physical computing systems and explain them.*

Exercise 102 *Explain the notion of information associated with levels of abstraction of a computational system.*

Exercise 103 *Explain the standard accounts of the physical Church-Turing Thesis.*

Exercise 104 *What is intended by functionality of a computational system and which elements contribute to it?*

Exercise 105 *What is intended by usability of a computational system and which elements contribute to it?*

Exercise 106 *What is intended by efficiency of a computational system and which elements contribute to it?*

Exercise 107 *Explain the linguistic account of the physical Church-Turing Thesis and how it differs from the standard account.*

11 Specification and Implementation

Summary

In Chapter 10 we investigated physical computing systems in terms of functionality, efficiency, and usability. This analysis has highlighted the need to understand the relation between specification and implementation of computing systems, and to clarify the problem of miscomputation. These two topics are considered in this chapter.

11.1 The Debate on Implementation

While formal verification has progressed with technical and industrial applications on compilers, processors, and large-scale web-based platforms,[1] the debate has continued—also in recent times—concerning the conceptual definition of its tasks and scope. On the one hand, the interpretation of programs as proofs allows one to reason about real computational systems as formal entities. On the other hand, critical positions on the role of formal methods underline their limitation in expressing some of the physical aspects inherent in implemented programs, such as their exposure to unpredictable interactions with the environment.

A first argument in this direction revolves around the issue of computational complexity of implemented software and our ability to survey programs. Recently, and as illustrated by the examples in the previous chapter, this aspect has received renewed attention with the huge improvements obtained through mechanical checking by satisfiability constraints solvers, proof checkers, and evolutions of Hoare-style logics. Accordingly, the

---

[1] See respectively the Coq verified C Compiler at http://compcert.inria.fr/index.html; the verification of microprocessors at IBM, see Paruthi (2010); Schubert (2015); and the Infer static analysis tool to detect bugs used on the Facebook platform, see Calcagno and Distefano (2011).

On the Foundations of Computing. Giuseppe Primiero, Oxford University Press (2020). © Giuseppe Primiero.
DOI: 10.1093/oso/9780198835646.001.0001

philosophical debate has focused on how these methods reduce error risks, and how they make the theoretical principle of program verification possible in practice,[2] thus recounting the objections of more sceptical approaches.

A second important position concerns the empirical and physical nature of computing, i.e. the relation between formal correctness and physical reliability investigated in Chapter 10. This aspect has hinged upon the *a priori/a posteriori* distinction, and the requirements of physical systems to qualify as computational.[3] From a philosophical standpoint, this problem has been often cast as one of software ontology, with different interpretations: software can be understood as patterns that causally affect machines when physically embodied;[4] software collects artefacts of an abstract, intensional, mind-dependent nature;[5] or software is defined in terms of data encapsulation of abstract structures, embedded with interaction patterns.[6] In all cases, most literature stresses the dual, abstract, and physical nature of programs.[7] More nuanced software-engineering approaches to the ontology of software have distinguished between different kinds of artefacts (code, program, software system, and software product), and described the ways these are interconnected in the context of a software engineering process.[8] The latest and more complete view on the ontology of computational artefacts distinguish functional from structural properties, where the former ones require design and the latter ones require manufacturing in view of implementation.[9]

In many respects, this debate reduces to the problem of explaining the link between the abstract and physical layers of computation, a relation which has traditionally been expressed in terms of *implementation*. We review in the following two main approaches to the topic of implementation and put forward a more detailed analysis.

The First Account

One early view on implementation focuses on the main relation between abstract and physical layers of computational systems as one between syntax and semantics:[10]

Definition 115 (Implementation as Semantic Interpretation) *I is an implementation of some syntactic domain A in medium M if and only if I is a semantic interpretation of a model of A.*

---

[2] See respectively Arkoudas and Bringsjord (2007) and Bringsjord (2015).

[3] See respectively Burge (1998) and Piccinini (2015).

[4] See Suber (1988).

[5] See Imrack (2012).

[6] See Colburn and Shute (2007).

[7] This is the case at least since Moor (1978). See also (Turner, 2018, ch.5) for the specific debate on the ontology of programs.

[8] See e.g. Wang et al. (2014).

[9] (Turner, 2018, ch.3).

[10] For this view, see Rapaport (1999, 2005).

Semantic Domain	is an implementation of	Syntactic Domain
computer program		algorithm
computational process		computer program
data structure		abstract data type

Figure 11.1 Examples of implementation relations in computer science according to Rapaport (1999)

Under this reading, the relation between the two main layers of a computational system involves

- a syntactic domain: the abstraction A, like an abstract data type;
- a semantic domain: the implementation I, the artefact or program that realizes A;
- a medium M, which can itself be abstract or physical and is typically rendered by a given programming language.

Examples of this relation are: an algorithm implemented by its formulation in a programming language; an abstract data type implemented by its formulation as a concrete data structure; the design pattern of a system given in terms of a finite state machine or unified modelling language implemented in Java classes. The implementation relations for the basic cases of interest in computer science are summarized in Figure 11.1

Note that this understanding of implementation relies on a *unique* interpretation (the semantic one) of the relation between abstract and physical layers. On the basis of our description of computational systems as construed by a layered ontology and associated epistemology (see Figures 10.2 and 10.3 in Chapter 10), this interpretation seems too reductive: a physical computer process is the semantic domain for the linguistic program; while at the same time the linguistic program (e.g. in Java) becomes the semantic domain to the corresponding algorithm. And so on. In other words, one is forced to accept a continuous shift of syntax and semantics across different levels, although this is not explicitly formulated in the notion of implementation presented. Implementation is required to close a semantic gap with its syntactic counterpart, and this gap is a different one at each ontological level. A second problem of this interpretation is that it does not account explicitly for incorrect implementations, nor for different types of miscomputation: the semantic account cannot describe errors happening at the different levels of the implementation relation between specification and physical realization, because it uses a unique relation at all the different levels.

The Second Account

To answer the correctness issue, another view insists on the normative role of the specification: the syntactic element fixes what the program *should* implement. In other words, the program will be correct only with respect to the specification, because it is the specification telling *what the program is supposed to do*. The problem is then to consider whether this interpretation guarantees the normativity of behaviour of the semantic counterpart. Moreover, this view describes implementation through the lens of the distinction between the abstract and the physical machine.[11]

The intensional level corresponds, roughly speaking, to the level of abstraction at which algorithms are considered as specifications of physical systems:

Definition 116 (Intensional Implementation) *An abstract machine is a specification of a physical one; it is taken to lay out the functional requirements of the physical device. The various components of the abstract machine must have corresponding physical ones for the latter to be a correct implementation of the former.*

Verification of the abstract model is possible mathematically, while the physical system as a realization of the former can only be verified empirically through testing. The latter is often imperfect and incomplete, with the specification guiding the formulation of test suites *sufficient* to guarantee a certain degree of correctness among the physical and the abstract layer.

A different way to consider implementation is by moving to the lower level of the physical machine and to consider it a correct implementation of an algorithm. This is the extensional notion of implementation:

Definition 117 (Extensional Implementation) *A physical machine correctly implements an abstract one if there is a mapping from the states of the former to those of the latter.*

This definition reformulates the simple mapping account from Chapter 10. Its main limitation is that an abstract model can be created a posteriori for any physical system (inducing the thesis that every system is a computational one). The difference with the intensional account is that correctness is not just guaranteed for operational uniform structures, instead it needs testing of each individual structure (thus inducing practical problems). Moreover, under this view it becomes hard to define in which way correctness fails, as there is no mapping to the intermediate level of language, or to the lowest level of physical actions.

A last sense in which implementation between an abstract and a physical machine can be defined is by referring to empirical practice:

Definition 118 (Empirical Implementation) *A physical machine implements an abstract one if an observer can identify a functional description of the former that matches the latter and which can be verified by checking that the theory of the abstract machine is correct and by testing and experimenting that the physical machine works acccordingly.*

This definition combines the previous two by requiring both the reconstruction of a functional mapping of the physical machine to the abstract one, and verification and testing to show correctness of its working according to specification.

The three definitions can be generalized as follows:[12]

---

[11] For the following definitions of intensional, extensional, and empirical implementation, see Turner (2012).

[12] See Turner and Angius (2017). See also (Turner, 2018, ch.4) for an illustration of the agreement between physical device and its corresponding logical machine in terms of the abstract–concrete interface.

Definition 119 (Implementation as Abstract–Concrete Relation) *An implementation I is the relation between the specification (abstract) and the artefact (concrete).*

This definition easily accounts for several types of implementations: the data type of finite sets implemented in the data type of lists represents a *mathematical relationship* (or model) where the axioms for sets act as a specification of the artefact, which in this case is implemented by the medium of lists; implementation across languages can be seen in a similar vein, with the mathematical structure underlying the construct in one language being translated in the appropriate construct for the other language; similarly, when the medium of implementation is a physical device, e.g. an abstract stack is implemented as a physical one, with the former providing the correctness criteria for the physical device. Note how this relation is recast for correctness at a lower level of abstraction than formal correctness:

> The notion of agreement that we are looking for is the correspondence between the truth table that characterizes the digital circuit and the extensional physical behavior of the electronic device: if we run all combinations of the electronic device and make a table of its input/output behavior, this must be in 1–1 correspondence with the truth table that provides the semantics of the digital circuit. It is this that links the abstract and the physical. Of course, in practice this is impossible for all but the very simple devices. But even when complete there is no mathematical guarantee of correctness. Unlike the correctness between the functional and structural specifications, this is an empirical not mathematical notion of correctness that is to be tested by physical not mathematical means. This is the abstract-concrete interface. The functional specification is a mathematical object — a truth table description of the input/output behavior of the circuit. Indeed, together with its truth table semantics, the structural description is also a formal mathematical device whose input/output behavior we can reason about independently any physical realization in electronic or any other physical medium. In contrast, an electronic device is a concrete physical thing.[13]

According to this analysis, implementation requires an independent semantic account in order to formulate a notion of correctness. This view then fits well with the approach that requires correctness of a physical program to be independently verified from its mathematical structure in a setting like Hoare logic. This notion of correctness is a property of computational artefacts.[14] In the following we will present our understanding for this criterion of correctness for implementations.

11.2 Correct Implementations

A refined version of the abstract–concrete relation can offer a more fine-grained interpretation of implementation for the layered structure of physical computing systems

---

[13] (Turner, 2018, pp.42–3).
[14] See (Turner, 2018, ch.25).

Epistemological Construct		Ontological Domain
Problem	aimed by	Intention
Task	fulfilled by	Algorithm
Instruction	expressed by	Programming Language
Operation	performed by	Machine Code
Action	realized by	Electrical Charge

Figure 11.2 Elements of the epistemology–ontology pairs

presented in Chapter 10, and it can also explicitly refer to levels of miscomputation which are not directly reduced to the abstract specification. We return once again to the methodological tool offered by the levels of abstraction (LoA) in a computing system. Recall the ontological domain illustrated in Figure 10.2, the epistemological structure from Figure 10.3, and the associated LoAs from Figure 10.1 in Chapter 10 , which were introduced with an explanation of which type of information each provides. Recall also that the epistemology provides the structure to understand the behaviour of the ontology. Let us then consider pairs composed by an epistemological construct and an ontological domain defining each LoA. Each pair is defined by an appropriate relation: an electrical charge realizes an action; machine code performs operations; programming language constructs express instructions; an algorithm fulfils a task; an intention aims at a problem; see Figure 11.2.

In this structure, an implementation is then the relation of instantiation that each pair of epistemological construct and ontological domain has with the pair above it, or with another pair at the same level, e.g. to account for implementation across languages (which is in fact still an implementation of the same algorithm in a different language). For example, the pair $\langle Problem, Intention \rangle$ is implemented in the pair $\langle Task, Algorithm \rangle$, and this in turn in the pair $\langle Instruction, PL \rangle$. Let us provide a general definition of such a schema:

Definition 120 (Implementation as Epistemology–Ontology Relation) *An implementation I is a relation of instantiation between pairs composed by an epistemological construct E and an ontological domain O of a computational artefact. A physical computing system is one which satisfies all the required levels.*

Now a correct physical computational artefact can be specified as correctness at each level, starting from the functional level:

Definition 121 (Functional Correctness) *A physical computing artefact t is functionally correct if and only if the functionalities of its intended specification S are displayed with the intended efficiency when t is used in the appropriate way.*

Here the interest is in the satisfaction of intended functionalities, which assumes efficiency and usability. Note that the notion of functional correctness can be further qualified in quantitative terms: a functionally *efficient* computational artefact t can be defined in view of the satisfaction of a minimal subset of functions required by its intended specification S; a functionally *optimal* computational artefact t can be given in terms of satisfaction of the maximal subset of functions required by (and still preserving) its intended specification S.[15]

At the lower level, we can offer a notion of correctness related to the algorithmic procedure of the artefact:

Definition 122 (Procedural Correctness) *A physical computing artefact t is procedurally correct if and only if it displays correctly the functionalities intended by the underlying algorithm A.*

Note that in this case the relation of correctness is determined as a relation between algorithm and program, and it assumes functional correctness. Procedural correctness as well can be further qualified in quantitative terms: a procedurally *efficient* computational artefact t correctly displays the functionalities with a minimal use of the logical resources available to the underlying algorithm A; a procedurally *optimal* computational artefact t correctly displays the functionalities with a maximal use of logical resources available to the underlying algorithm A (while not hindering functionalities of other algorithms in the same system).

Finally, at the lowest level:

Definition 123 (Executional Correctness) *A physical computing artefact t is execution-ally correct if and only if its underlying program runs correctly on the architecture on which it is installed.*

In this formulation, correctness refers to the relation between the structure of the program and its physical realization on a given architecture, assuming functional and pro-cedural correctness. Executional correctness qualified in quantitative terms is as follows: an executionally *efficient* computational artefact t correctly runs on the architecture on which it is installed with a minimal use of the physical resources available to the system; an executionally *optimal* computational artefact t correctly runs on the architecture on which it is installed with a maximal ues of the physical resources available to the system (while not hindering other functionalities of the system).

The collation of the three notions of correctness gives us a general notion of correct physical computing system:

Definition 124 (Correct Physical Computing System) *A correct physical computing system is one which presents correct implementations at all the required levels.*

A correct instantiation at each level is required for a well-functioning computing system: the intention of the designer to solve a problem is implemented into an algorithm to resolve the corresponding task; the algorithmic task is implemented in a program to express the related set of instructions with appropriate data; this in turn is implemented into low-level language operations for machine code; and finally these control actions are implemented by electrical charges.

---

[15] These notions are defined and formally developed for a logic of design in Primiero (2019a).

A mismatching implementation is then a cause for error in the computational system, with as many types of miscomputation possible as there are levels of abstraction. When considering an informational account of computation as by Definition 107, the problem of programs that 'can fail at many different levels'[16] can be approximated by an analysis of *levels of information failure*. When the instatiation across pairs is rendered incorrectly, a failure at a given level of abstraction occurs. In Section 11.3, we recollect a theory of miscomputation which is designed to account for the various levels of abstraction of a computing system.

11.3 Miscomputation

A correct system will present a series of matching implementations across the pairs composed by an ontological domain and an epistemological construct at each level of abstraction. Matching implementations at all levels express a correct procedure being linguistically formulated and physically executed to fullfil an algorithm solving the intended problem. A first sense in which a computational system might be said to be incorrect is the wrong coupling of intention and algorithm, i.e. where an inappropriate algorithm is chosen — though possibly well executed and therefore syntactically correct — but it does not solve the intended problem. This is an error of design, but it generates a possibly correct computational system for an output different than the one sought. We shall not further explore this case, as it defines a conceptual limitation which cannot be addressed but in terms of a new design.

A second sense of incorrect physical computational system is instead a malfunctioning, i.e. an execution error which makes the procedure at one of its levels of abstraction incorrect for the problem at hand. When executed correctly at all levels, the algorithm offers instead an appropriate solution to the problem at hand.[17] Within this second type of incorrect formulation of a computational system, the problem may lie in each of the levels identified above. We can distinguish conceptual and material categories of miscomputations, split into three families; see also Figure 11.3:

1. Conceptual errors reflect breaching of validity conditions: in this family,

   - mistakes represent errors at the Functional System Level, where the problem description can be wrong in terms of contradicting requirements;

   - failures occur at the Design System Level, where the problem is expressed in terms of an invalid design;

   - and slips occur at Algorithm Design Level, where algorithmic constructions use invalid routines.

---

[16] See (Cantwell Smith, 1996, p.813).

[17] For a formal translation of the distinction between the first and second sense of correct computational system, see Primiero (2014).

Error	Conceptual	Material	Performable
Mistake	Problem Description	Problem Design	
Failure	Procedure Definition	Procedure Construction	
Slip	Algorithm Design	Algorithm Implementation	Algorithm Execution

Figure 11.3 Families of computational errors, Primiero (2014)

LoA	Type of Error
Functional System Level	Contradicting Requirements
Design System Level	Invalid/Incomplete Design
Algorithm Design Level	Invalid/Incorrect Routine
Algorithm Implementation Level	Syntax/Semantic/Logic Error
Algorithm Execution Level	Wrong/Failing Hardware

Figure 11.4 Levels of abstraction and associated errors, Fresco and Primiero (2013)

2. Material errors reflect breaching of correctness conditions: in this family,

- mistakes at the Functional Specification Level refer to incomplete procedural definitions;

- failures at the Algorithm Design Level refer to incorrect design of the algorithm

- and slips express incorrect translation at the Algorithm Implementation Level by the introduction of a syntax, semantic, or logic error.

3. Performable errors reflect breaching of physical conditions and happen at the Algorithm Execution Level because of physical failure or inadequate physical support.

We summarize these types of error in Figure 11.4. The distinction between mistakes, failures, and slips is functional to the definition of intensional and extensional errors which we have identified as a limitation of the mechanistic interpretation of physical computing in Section 10.1. Simply put, a mistake will induce the computational system to produce something else than the intended specification, or no output at all in the worst case. On the other hand, failures occur at the algorithm and software level and they can manifest themselves in the form of unintended or undesired functionalities.

This difference is reflected further in the distinction between the notions of misfunctioning and disfunctioning computational process.[18] These notions are defined for physical computing artefacts analysed at all levels of abstraction:

Definition 125 (Disfunction) *A physical computing artefact t disfunctions if and only if it is less reliable or effective in performing its function F than one justifiably expects for artefacts of its type T.*

Definition 126 (Misfunction) *A physical computing artefact t misfunctions if and only if it satisfies the following three conditions:*

---

[18] This distinction and the definitions that follow are introduced and explored at length in Floridi et al. (2015).

1. *using t produces some specific side effects e of type E;*

2. *because of e, one has reason not to use t; and*

3. *other programs of the same type do not produce the same side effects of type E.*

While strictly speaking disfunctions reflect mistakes and misfunctions reflect failures, in a weaker sense also performance error can be understood as misfunctions, as efficiency criteria are often formulated as specifications: therefore, processes that do not match the intended performance may be dubbed to display unintended side effects in terms of higher complexity than planned or expected.

In view of the conceptual machinery here introduced, it becomes possible to address the notions of correctness and validity for physical computational artefacts. This means to make explicit the correctness notion for physical computing systems given in Definition 124 in terms of the properties of functionality, usability, and efficiency introduced above. This is the aim of Chapter 12.

Exercises

Exercise 108 *Explain the notion of implementation as semantic interpretation and provide some examples.*

Exercise 109 *Explain the different notions of intensional, extensional, and emprirical implementation as an abstract–concrete relation and provide some examples.*

Exercise 110 *Explain the understanding of implementation as a relation between pairs of ontological domains and corresponding epistemological constructs in a computational system.*

Exercise 111 *Explain how correctness can be defined at different levels of abstraction.*

Exercise 112 *Clarify the different errors that can occur in a computational system.*

Exercise 113 *Give an explanation of the difference between disfunctions and misfunctions.*

12 Computing as an Engineering Discipline

Summary

In Chapter 11 we investigated the notion of correct physical computing systems at several levels of abstraction, and defined forms of malfunctioning. In the present chapter, we reconsider the engineering foundation of computing, overview the related debate, and formulate a notion of physical computational validity.

12.1 Software Engineering

From the laws of computing evolution to the increasing reliance on implementation for the correctness of computational systems, it clearly appears how computing has progressively become more significantly affected by software than by hardware.

Since the early days of computing, programming was aided by annotations for correctness control. This practice evolved from flowcharts to mathematical formulas, reaching an important turning point with the design of the ALGOL language in 1958 and 1960. ALGOL represents an iconic language, in that it endorses at the software level the entire problem of correctness, but it is also representative of the status of software in those decades. ALGOL was born from the spirit of a community willing to construe a language around the formal properties of logical notation and structure. But its latest formulation ALGOL68 divided the community, when an alternative design suggested by Adriaan van Wijngaarden was considered too ambitious, implementing too many unnecessary features, with a difficult notation:[1] this meant integrating a view of the programmer's task that made it obsolete as a programming tool of reliable complex

---

[1] See (Haigh, 2010, p.5).

On the Foundations of Computing. Giuseppe Primiero, Oxford University Press (2020). © Giuseppe Primiero.
DOI: 10.1093/oso/9780198835646.001.0001

software systems.[2] The history of ALGOL is in this sense emblematic of the significant switch from hardware to software as the most essential element of computing, a process occurring throughout the 1960s with the formulations of Moore's prediction in 1965 and Amdhal's Law in 1967 representing the first attempts at identifying computational growth from a software perspective. This direction was obviously driven by an increasing diffusion of cheaper hardware, leading to an extensive development of more complex software systems, in turn requiring increasingly sophisticated programming competencies, methods, and strategies.[3] More elements intervened in this process: the transition from scientific to the business applications; the beginning of institutionalized training and education for computing professionals, with the birth of the first Computer Science Departments in Universities;[4] the transformation from batch processing to time-sharing systems, including multi-processing and concurrent programming, with the DEC PDP-1 first and the IBM 360/67 afterwards.[5]

When recollecting the beginning of his career as a programmer in the early 1950s, Dijkstra reconsiders his choice of leaving theoretical physics to become a programmer:

> A programmer? But was that a respectable profession? After all what was programming? Where was the sound body of knowledge that could support it as an intellectually respectable discipline?[6]

According to Dijkstra, the uncertainty about the discipline was still at that point significantly contrasted with the role of hardware specialists, who could speak in terms of their engineering knowledge and capabilities. And while the early opinion about the programmers' role was constrained to processes' optimization, the availability of more complex and powerful machines instead of making programming unproblematic revealed the 'software crisis'.[7] Dijkstra identifies a major issue in the development of the third generation of computers, where the ratio between price and performance meant that their design was not sound for their programming. At the beginning of the 1970s, Dijkstra's vision was that by the end of the decade computing professionals would be able to design and implement systems without straining the programming abilities, and at only a fraction of the costs of the time. Typically for his background, he would add that these systems

---

[2] See Dijkstra et al. (1970). For a history of ALGOL, see e.g. Lindsey (1996); de Beer (2006); Nofre (2010).

[3] For a history of the programming community from the 1950s to the 1970s, see Ensmenger (2010).

[4] The earliest degree in computer science was established at the University of Cambridge in 1953; the first PhD graduate was Richard Wexelblat at the University of Pennsylvania, Graduate Group in Computer and Information Sciences, founded in 1959 in the Electrical Engineering Department; the first Computer Science Department in the United States was established at Purdue University in October 1962, see https://cacm.acm.org/blogs/blog-cacm/159591-who-earned-first-computer-science-ph-d/fulltext.

[5] For a taxonomy of operating systems of the 1950s and 1960s classifying them on the basis of automatic programming, batch-processing, integrated systems, command and control systems, process control systems, teleprocessing, and time-sharing systems, see Bullynck (2018).

[6] (Dijkstra, 1972, pp.859–60).

[7] See (Dijkstra, 1972, p.860).

would be correct. In fact, correct design and implementation of complex and powerful implemented systems were the dream of computing professionals of the 1960s and 1970s.

In 1968 another crucial event for the evolution of the computing discipline, and in particular of software, occurred: a NATO-sponsored conference on *software engineering*. The agenda of this meeting was anticipated in spirit by the troubles mentioned above for ALGOL68, concerning its structure, features, and notations,[8] and the perception of the software crisis was formulated openly for the first time. Aims of the conference initiated by a Study Group on Computer Science of the NATO Science Committee were:[9]

- achieving sufficient reliability in integrated data systems;
- meeting schedules and specifications on large software projects;
- the education of software engineers;
- whether software should be priced separately from hardware.

In this context, the definition of software engineering was based on the nature of software projects as illustrated in Figures 12.1 and 12.2.

At these early stages, the structure of levels of abstraction between design and implementation was formalized in terms of a distinction between external and internal

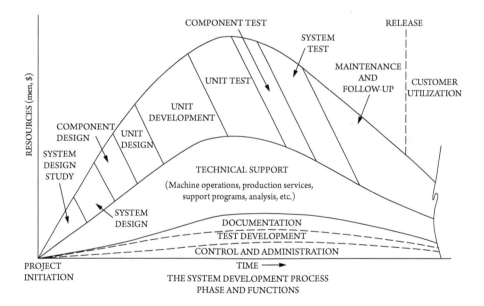

Figure 12.1 Software projects, (Naur and Randell, 1969, p.20)

---

[8] See Haigh (2010).
[9] See (Naur and Randell, 1969, p.3).

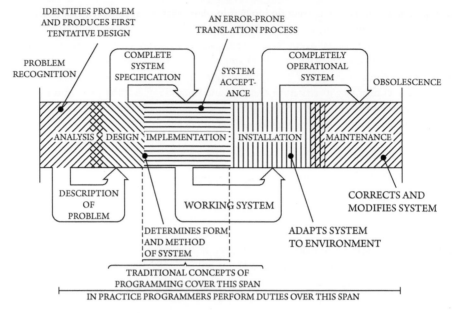

SCHEMATIC FOR THE TOTAL PROGRAMMING
(SOFTWARE SYSTEM-BUILDING) PROCESS

Figure 12.2 Software projects, (Naur and Randell, 1969, p.21)

specifications with a feedback from the latter to the former. The methodology was illustrated as comprising knowledge and understanding of what a program is, and the set of methods, procedures, and techniques by which it is developed.[10] The two phases were referred to as *design* and *production*. Program design was illustrated in terms of:

a. Specification of the complete hardware–software system.

b. Definition of the functions to be performed by the program.

c. Design and documentation of the master (overall) program plan.

d. Subdivision of the large system program into manageable program blocks.

e. Definition and documentation of interaction between program blocks.

f. Basic program subroutines defined and documented.

g. Detail design, coding, and documentation of each program block.

h. Design and documentation of test methods for each program block in parallel with step (g).

i. Compilation and hand check of each program block.

[10] See (Naur and Randell, 1969, pp.24–5).

j.   Simulation of each program block using test methods planned during the design of the program.

k.   Test and evaluation of the program blocks in the system.

l.   Integration of complete program in the system.

m.   Final load testing of the complete software–hardware package to see that the program meets all of its design requirements.

This early qualification of the engineering process includes everything known today, from requirements elicitation to system testing. Along with the program development methodology, other crucial aspects involved by this process are the typology of system designers (programmers) and the interaction of software with the host system. This understanding of program design encompassed processes of correctness checks:

> Dijkstra: 'Whether the correctness of a piece of software can be guaranteed or not depends greatly on the structure of the thing made. This means that the ability to convince users, or yourself, that the product is good, is closely intertwined with the design process itself.'[11]

Correctness was here associated not only with flowchart design for the understanding of the problem but also with an iterative process of coding and redrawing of the charts until *you feel [you obtain] the correct solution*.[12] The problem with this process starts as soon as the system to be built is big enough to require 200 people writing code and iteration through code and specifications is cut short by production deadlines. In the face of this reality, a certain position starts to emerge among programmers that an iterative workflow of this kind stemming from well-defined specification is in fact impossible. Two schools of thought (not necessarily in full opposition) seem to emerge from the attempt at clarifying the problem of software design:

- on the one hand, the group of participants who stressed the mathematical nature of software and the need for powerful mathematics, concise notation, and linguistic constructs to express essential structure and relationships general to programming rather than specific languages;[13]

- on the other hand, those who would stress accessibility, modularity, documentability, and user experience.[14]

---

[11] (Naur and Randell, 1969, p.31).

[12] Comment by Kinslow, (Naur and Randell, 1969, p.32).

[13] Among proponents of this approach are Perlis, Bauer, and Kolence; see (Naur and Randell, 1969, pp.37–8).

[14] Among proponents of this approach are Perlis, Letellier, Smith, Gillette, and Babcock; see (Naur and Randell, 1969, pp.38–9).

One way or another, reliability and logical completeness with respect to sets of operations remained crucial tasks of the ideal design process. Discussed design strategies included:

- sequencing (intended as the order in which to do things, distinguished already in top-down and bottom-up approaches);
- structuring (intended as the resulting structure of software design with appropriate properties, e.g. coherence, decomposability in view of relationships and transformations over data structures, redundancy, and of course for some provability);
- monitoring and simulation (intended as testing of partial design through high-level languages for function description, data structure description, input/output patterns, online display for variables and functional relations);
- high-level languages for abstraction from hardware and related compilers design;
- and the need for notations for communication during the design process.

Program production, on the other hand, referred to *'the total task of designing, implementing, delivering, maintaining, etc., a software system'*.[15] The first factor analysed in this context was the growth rate of the size of software systems by lines of codes, which raised from 5.000 in 1954 with the IBM 650, to 2.000.000 in 1967 with the IBM 360.[16] Today, by comparison, among the largest software systems are the Google web services, which include more than 2 billion LoC, and the Human Genome Project, which counts over 3 billion.[17]

A second, strictly connected factor was the relation between system size and number of developers:

> Perlis: 'We must learn how to build software systems with hundreds, possibly thousands of people. It is not going to be easy, but it is quite clear that when one deals with a system beyond a certain level of complexity, e.g. IBM's TSS/360, regardless of whether well designed or poorly designed, its size grows, and the sequence of changes that one wishes to make on it can be implemented in any reasonable way only by a large body of people, each of whom does a mole's job.'[18]

Additional factors mentioned were the number of different hardware configurations which the software must accommodate, and the range of input error conditions which the system must handle gracefully. Obviously, a problem often mentioned concerned the increasing economic costs associated with large projects.

Also production focused on the issue of reliability, with what can be considered the origin of the RAS system by IBM, illustrated in Chapter 10: the need to have no

---

[15] (Naur and Randell, 1969, pp.64–5).
[16] See (Naur and Randell, 1969, p.66).
[17] https://informationisbeautiful.net/visualizations/million-lines-of-code/.
[18] (Naur and Randell, 1969, p.68).

'more than two hours system downtime (both software and hardware) in 40 years',[19] and to establish how much error can be considered acceptable in a system containing one million instructions.[20] Management aspects of the production phase discussed were, among others:

- planning;
- the well-known over-run in size (man-months);[21]
- personnel efficiency, ability, morale, supervision;
- communication and control over progress made (including documentation);
- tools: for scheduling, debugging, access prototypes and common working space, compiling, testing, building, customizing, replicating.

In this context, the software crisis was formulated in several ways:[22] as the widening gap between ambitions and achievements, or promises to users and performances achieved by software (by David and Fraser); as the problem of producing systems on schedule and to specification (Gillette); as the possibility of failure in safety-critical systems and in industry (Randell and Opler); as the relevance of bad software to society (Perlis); as the demands of systems beyond capabilities, theory and design methods (Kolence). The causes, as well, were multiple:[23] the gap between working in theory and functionality in practice and economic pressure (Buxton); the telescoping of research, development and production (David and Fraser); the refusal of industry to re-engineer previous models, its inability to allow personnel to accumulate applicable experience, and emotional management (McClure); incomplete understanding of the design process (Kolence); users who are not aware of the requirements of desired features (Hastings). Among the possible partial solutions there were:[24] different ways of announcing and promise projects (Opler); evolutionary design and production (Kinslow); change of practice in software design and production (David and Fraser); development of existing techniques (Gill, Perlis).

Twenty years after the NATO Conference, the situation did not appear to have improved to many practitioners, and they gave up any hope of seeing the kind of improvement that guided hardware evolution:

> Not only are there no silver bullets now in view, the very nature of software makes it unlikely that there will be any — no inventions that will do for software productivity, reliability, and simplicity what electronics, transistors, and large-scale integration did for computer hardware. We cannot expect ever to see twofold gains every two years.[25]

---

[19] (Naur and Randell, 1969, p.70).
[20] To this observation by Opler, Perlis provocatively replies *'Seven'*.
[21] A topic notoriously discussed at length by Frederick Brooks; see Brooks (1975).
[22] See (Naur and Randell, 1969, p.120 and ff.).
[23] See (Naur and Randell, 1969, p.122 and ff.).
[24] See (Naur and Randell, 1969, p.124 and ff.).
[25] (Brooks, 1987, p.10).

Analyses pointed to the slow speed of software progress, induced by essential difficulties (complexity, conformity, changeability, and invisibility) as well as accidental ones (high-level languages, time-sharing, and unified programming environments).

In this context, the slow progression of software evolution was necessary for a serious engineering science, and as such the management and control of the crucial transitions between phases of a software project (planning, design, code, test planning, and evaluation of test results) was to be aided by both formal and informal reasoning, as well as testing and its strategies requiring complete documentation, design inspection, and review; test-based quality assurance; error removal and defensive programming, and the direct and still foundational role of formal methods and proof.[26] The methodology of programming remained at the basis of the practitioners' approach, with the gap between theory and practice one of the main indications of the growing importance – but also problematic nature – of the new discipline.

12.2 The Debate

The era of integrated circuits, in particular the exponential rise expressed by Moore's Law, and similar progress in other computing technologies highlight a novel approach to the discipline, a way of characterizing its foundations which is much closer to engineering than the mathematical approach expressed by principles like the Church-Turing Thesis. The above analysis of the principles of software engineering during the 1960s and in the following decades has illustrated this conceptual shift. Accordingly, also the debate around the nature of the discipline at large changed, with a progressively stronger stress on the physical and material nature of computing, as opposed to its mathematical foundation. In 1967, Newell, Perlis, and Simon argued that

> Computer science is the science of computers. The phenomena surrounding computers are varied, complex, rich.[27]

As computing machinery can be taken to be an engineering artefact, it seems that this definition qualifies computer science in the first place as an engineering science. In 1968, the ACM Curriculum argued that computer science embraces elements of both science and engineering. In his Turing Award lecture in 1968,[28] Hamming posited the technology of the computing machine at the core of the discipline: the theoretical question whether something can be done is considered less important for the discipline than finding a *cost-effective* way of building it. Abstract computation is therefore less central than its practical implementations.

---

[26] See Hoare (1996).
[27] (Newell et al., 1967, pp.1373–4).
[28] Hamming (1968).

This trend was followed for the following 30 years, in particular in the comparison with the sciences: was the engineering nature of computing sufficiently scientific so as to be qualified as part of the experimental paradigm of science (and thus also justifying important scientific investments)? Peter Denning stressed the engineering nature of computer science as a discipline in referring to the need for efficiency:

> The body of knowledge of computing is frequently described as the systematic study of algorithmic processes that describe and transform information: their theory, analysis, design, efficiency, implementation, and application. The fundamental question underlying all of computing is, What can be (efficiently) automated?[29]

In his 1993 Turing Award lecture, Hartmanis dwells on the complex nature of computer science: it is characterized as a new species among the sciences, in such that it deals with the creation and processing of information, and as such it is not directly restrained and governed by the laws of physics. A specific aspect is then represented by the various orders of magnitude involved by the several levels of abstraction needed to deal with all the aspects related to information, from bits to programs to systems, and the precision required to do so. The power and scope of the computing paradigm is given by the universal nature of the machine as devised by Turing (and generalized, among others, by Hartmanis himself, who developed the multi-tape version). But computing theories do not need to be constrained by observation, in that reflecting a more creative, engineering spirit:

> theories do not compete with each other as to which better explains the fundamental nature of information. Nor are new theories developed to reconcile theory with experimental results that reveal unexplained anomalies or new, unexpected phenomena, as in physics. In computer science there is no history of critical experiments that decide between the validity of various theories, as there are in physical sciences.[30]

Loui strenghtens this characterization of computing as a discipline whose aim is essentially linked to usefulness, efficiency, and economic constraints. As such, it qualifies as a *'new species of engineering'*.[31] Albeit rooted in the mathematical foundation of computing, the enginneering nature of the discipline cannot be separated from its more physical aspects. McKee recalls his experience of the previous 15 years to claim that

> people in computer science had different goals and methodologies than scientists in traditional fields.[32]

On this basis, he suggests replacing the term 'computer science' with 'computics': in this discipline, data is only produced by the computer, according to its design, and theory and

---

[29] (Denning, 2000, p.405).
[30] (Hartmanis, 1994, p.40).
[31] See (Loui, 1995, p.31).
[32] (McKee, 1995, p.136).

experiment are only weakly linked to each other. Computics is an engineering science because its aim is essentially to minimize costs and maximize utility in the process of machine design. The issue of the name of the discipline comes back one year later, in the lecture for the Newell Award delivered by Frederick Brooks. In particular, by criticizing the characterization of the discipline as a science, Brooks proposes again the relation to engineering:

> Perhaps the most pertinent distinction is that between scientific and engineering disciplines. That distinction lies not so much in the activities of the practitioners as in their purposes. [...] the scientist builds in order to study; the engineer studies in order to build.[33]

The practitioners of the discipline are concerned with *'making things, be they computers, algorithms, or software systems'* and as such computer science *'is in fact not a science but a synthetic, an engineering, discipline'*. Despite the fact that most of computing artefacts are abstract objects, these are used by people to enrich human living: the role of the practitioner of the discipline is like that of a toolsmith in that the success relies in what its tools achieve when used. Naming the discipline as a science has, according to Brookes, many negative effects: it seems to put the role of science higher than that of engineering; it makes the invention and publication of computers, algorithms, and languages an end in itself (rather than a tool to make things happen); it allows the users and their needs to be forgotten; finally, it allows the mathematical aspect of the discipline to be considered more important than its practical side.

The hybrid nature of the computing discipline *'as a third paradigm of science, joining theory and experimentation'*[34] is qualified by three major elements:

1. *Theory*: building conceptual frameworks and notations for understanding relationships among objects in a domain and the logical consequences of axioms and laws.

2. *Experimentation*: exploring models of systems and architectures within given application domains and testing whether those models can predict new behaviours accurately.

3. *Design*: constructing computer systems that support work in given organizations or application domains.

The transformation of the computing discipline from a mathematical to an engineering discipline can be summarized in terms of the following requirements:[35]

- practical implementation alongside abstract ideas;
- demonstration alongside proof and experiments;

---

[33] (Brooks, 1996, p.62).

[34] (Denning, 2000, p.408).

[35] For an especially detailed analysis of the views on computing as an engineering science, see Tedre (2009).

- questions about what can be realized alongside questions about what exists in reality;
- art together with science;
- technology and applied science alongside basic research;
- problem-solving and engineering methodology combined with theory (e.g. properties of algorithms), decision-making, and process optimization;
- design of abstract machines along with design of concrete, resource-bounded artefacts.

This characterization of computing as an engineering discipline stresses functionality, value of the working artefacts, impact on reality, processes, and products, implementation of abstract ideas, descriptive and normative knowledge, partiality.[36] Denning defines computing as *'the science of information processes and their interactions with the world'*, a definition which is formulated on the basis of an understanding of a set of principles *'that extended beyond its original mathematical foundations to include computational science, systems, engineering, and design'*.[37] In this definition science, engineering, and mathematics are all components of the discipline and the scientific paradigm of forming hypotheses and testing them through experiments is applied to information processes. The characteristic of computing which makes it distinct from other natural sciences is that its object of study is information processes *'both artificial and natural'*.[38]

After reaching its peak of exposure and agreement, this characterization based on engineering began to be criticized for being too limited, in particular because it does not reflect the complex aspects related to influence in human-made processes, the role of experimentation, and the relation with other disciplines.[39] It will be precisely by reinforcing these additional aspects that a new understanding of computer science was emerging, related to its experimental nature and the use of computerized methods to perform experiments in other sciences. We shall explore further the experimental nature of computing in Part III of this volume. Before that, let us return to the crucial role of information in the definition of the discipline, how it is strongly connected with its methodology and which notion of validity is available under the engineering interpretation of computing.

12.3 Physical Computational Validity

Starting in the 1960s and for several decades thereafter, the model of relevance for computation has progressively moved from an abstract mathematical one to the physical

---

[36] See (Tedre, 2009, p.1655).
[37] (Denning, 2005, p.27).
[38] (Denning, 2005, p.28).
[39] See e.g. Denning (2000).

realization of input/output information transmission. In this conceptual process, the evolution of the essential elements of the engineering foundation of computing (code, architecture, hardware), the associated programming practices required (physical wiring, encoding, read-only instructions, storing along with execution, editing) as well as the methodological aspects (design, implementation, delivery, maintainance) have all played essential roles.

The notion of computational correctness for physical computing systems presented in Definition 124, Section 11.2, has complemented the logical notion of correct inference inherited from the mathematical foundation, on the basis of two main conceptual shifts:

1. an interpretation which moves away from models, and it is defined in terms of physical implementation;

2. an interpretation of implementation in terms of several levels: functional, procedural, and executional.

This allows one to identify limitations to correctness due to the physicality of computational artefacts, as induced by this layered structure. It moreover makes it possible to reconsider a notion of validity appropriate for physical computational systems. This notion must be formulated in the context of a general understanding of the discipline as a science of information, in line with its object being the study of artefacts manipulating input/output relations. Attempts on those lines are not new in the literature, and they are certainly natural given the main aim and objective of computing is centred around information representation, processing, and use. Already in 1989, the algorithmic nature of computing was formulated by Denning on the basis of information processing:

> The discipline of computing is the systematic study of algorithmic processes that describe and transform information: their theory, analysis, design, efficiency, implementation, and application.[40]

From this definition, it is clear how information can denote both structure and content of the computational processes. A general account of computing as a science of information as the one offered above can be improved relying on the formulation of computational artefacts preserving all the relevant levels of abstraction. Each such level expresses a way in which information is defined, structured, processed, represented, and extracted. Recall from Section 7.4 that we have identified three main LoAs, each with a given domain associated, where for each layer a specific definition of information is offered; see Figure 7.11. Recall moreover from Section 10.1 that these categories are refined for computational artefacts in their physical implementation. In particular: at the Functional Specification Level, abstract information is the content of the designer intention given in terms of requirement specification; at the Algorithm Design Level, it is still abstract information the content of rules description for the intended specification;

---

[40] (Denning et al., 1989, p.12).

at the Algorithm Implementation Level, instructional information is the content of the high- and low-level language translation of the algorithm; finally, at the Algorithm Execution Level, operational information is the content of the execution of the algorithm in hardware. Through this description, we are able to distinguish information content from information structures of computing. This duality has been reformulated in terms of the distinction between a domain of objects (or ontology) with an associated structure of control (or epistemology); see Figure 11.2. The ontology–epistemology duality is therefore crucial to offering a revised definition of computing as a science of information:[41]

Definition 127 (Computing as Science of Information) *Computing is the systematic study of the ontologies and epistemology of information structures.*

With the term 'systematic study' we encompass the scientific methodology relying on the formulation of hypotheses, deductive reasoning, and experimental testing which we have seen has been endorsed in previous formulations of computing as a science of information. With the reference to 'information structures' we aim at including all areas of interest, from theory, analysis, design, efficiency, to implementations and applications. Finally, this definition stresses the methodological relevance of distinguishing information as the domain of objects (bits, data, syntactic constructs, algorithms, intention) from information as the structure of control (gates, data structures, semantic constructs, rules, problem domain). In philosophical terminology, these express respectively the *know-that* vs. the *know-how* of computing.

The issue of validity of physical computational systems can now be recast as the problem of establishing validity for the ontologies and epistemology of information structures. In the mathematical model, the relation between specification and computation is rendered by fixed logical relations, formulated in the Principles 5 and 6 of Algorithmic Dependence and Enconding Independence from Chapter 7. They reflect an essential reliance on the algorithmic description and its lingustic expression, but show a full abstraction from the procedural level of algorithm execution. The move from the mathematical to the engineering understanding of computing requires, instead, a strong reliance on the given physical architecture according to the requirements of the formal specification for data representation, code optimization, and process execution. The new understanding of computing makes the case for a stronger role of the physical properties of implementation.

First, the algorithmic description is no longer a sufficient representation of the principle of computable structure, and one requires reference to implementation in a given physical architecture realizing the algorithmic description:

Principle 12 (Architectural Dependence) *A physical computation is the implementation in a given architecture of the linguistic encoding of an algorithmic process.*

Second, such implementation is not irrelevant to the input/output relation, but it rather becomes an essential element in defining which processes can be defined as compu-

[41] See also (Primiero, 2016, p.122).

tational and according to which limits. In other words, the behaviour on the input/output relation is essentially dependent on the physical process:

Principle 13 (Behavioural Dependence) *A physical computation may behave differently depending on the architecture's properties.*

One could stress that the success of the modern code paradigm and of the von Neumann architecture have been in line with the conceptual principles highlighted by the mathematical foundation. But it is also clear that the evolution of computational principles according to the physical embodiment (miniaturization, parallelism, compatibility) and how their properties affect the related behaviour is a novelty of the engineering foundation. Along these lines, the notion of validity for the ontologies of physical computing can be abstractly defined as follows:

Definition 128 (Valid Physical Computation — Ontological Version) *Given an intended specification S, we say that physical artefact t processing information linguistically expressed by an algorithm A implemented in a program P using resources $\mathcal{N}$, running on an architecture $\mathcal{A}$ is physically valid, denoted*

$$(\mathcal{N}, P(A)) \vDash_{\mathcal{A}} S$$

if and only if t is a physically correct computing system producing a state satisfying S.

This definition collates the various levels of abstraction with correctness and specification satisfaction. Note that an entirely matching definition can be offered which focuses solely on the epistemological level:

Definition 129 (Valid Physical Computation — Epistemological Version) *Given a problem $\mathcal{P}$, we say that a step-by-step task t linguistically implemented by a set of instructions I using resources $\mathcal{N}$ and executed through an action $\mathcal{A}$ is physically valid, denoted*

$$(\mathcal{N}, t(I)) \vDash_{\mathcal{A}} \mathcal{P}$$

if and only if

- *t displays the intended behaviour (functional correctness);*
- *t acts according to I (procedural correctness);*
- *$\mathcal{A}$ executes I (executional correctness);*

and t through $\mathcal{A}$ produces a state solving P.

While the ontologies and epistemology of computing have been largely explored in this section, the remaining element to explore and analyse is the scientific nature of their study, in terms of the revisitation of the hypothetical-deductive method known from other sciences. The question of the scientific value of computing relies on the clarification of its experimental method. This exploration is the task of Part III of this volume.

Exercises

Exercise 114 *Clarify the context of the 'software crisis', illustrate the scope of the problems considered, and illustrate what methodologies were devised to solve them.*

Exercise 115 *Which elements characterize Computing as an Engineering Discipline?*

Exercise 116 *Which are the most disguishing features of engineering with respect to mathematics?*

Exercise 117 *Clarify the notion of Computing as a Science of Information.*

Exercise 118 *Explain the meaning of the Architectural Dependence Principle.*

Exercise 119 *How does the notion of valid physical computation reflect the understanding of computing as a science of ontologies and epistemology of information structures?*

Exercise 120 *What are the essential requirements for the ontological version of validity for physical computation?*

Exercise 121 *In which way does the physical understanding of valid computation differ from its formal counterpart?*

Part III
The Experimental Foundation

13 Elements of Experimental Computing

Summary

In this chapter, we start by considering the third foundation: computing as an experimental science. We overview the origin of the term, the early positions on experimental computer science and introduce two basic notions of computational hypothesis and computational experiment essential to the understanding of computing as an experimental discipline.

13.1 Experimental Computer Science

Two documents are significant in identifying the beginnings of the trend emerging in the 1980s which recognizes computing as an experimental discipline.

The notion of *experimental computer science* can be traced back to the report of a workshop sponsored by the National Science Foundation (US), the so-called Feldman Report.[1] The starting observation made in the report was that, while most of the foundational results and manpower in computer science came from universities, two main factors were changing the landscape: first, the technological advances were requiring a more experimental approach to the discipline; second, the best academics were being attracted by the industry. The contribution of academics was detailed in areas like

(i) large management systems, with the development of time-sharing systems, multi-programming, virtual memory, and associated dynamic storage techniques;

---

[1] Feldman and Sutherland (1979).

On the Foundations of Computing. Giuseppe Primiero, Oxford University Press (2020). © Giuseppe Primiero.
DOI: 10.1093/oso/9780198835646.001.0001

(ii) office automation with the combination of graphics, document preparation, distributed databases; and

(iii) communication and networking technologies.

These examples were clearly chosen to stress the relevance of theoretical work on industrial applications. In this light, the discipline was characterized as *experimental computer science*.[2] In this report, the qualification of 'experimental' mainly referred to the role of computing in working and industrial environments, as opposed to the theoretical research going on mostly in universities. The experimental side of computing included *'advances in microelectronics, communications, and software technology'*.[3] The role of universities was recognized as foundational in the early development of the information processing sector. The development of new hardware techniques and the extensions of computing to new technologies were rightly foreseen as the beginning of a new critical era, in which resources and capabilities would flow from the theoretical to the experimental arena. A crucial aspect of the (new) experimental nature of computing was represented by the practice of testing and the high costs associated with its development and the required equipment: interestingly, these highlighted the relevance of a methodological shift in computing itself and the influence that the new discipline would exert (or be exerted on) by its use in other sciences, ranging from biology to natural language processing to algebra. Ultimately, the report was meant as a plea to avoid a fracture between theory and applications, by granting the necessary investments in the academic environment, as well as strengthening their relationship in terms of a reciprocal flow of money, equipment, and services.

The second document was a report following the biennial meeting of Computer Science Department Chairmen held at Snowbird, Utah in 1980, attended by 56 department heads or their representatives, and by six observers from industry and government. The so-called Snowbird Report[4] illustrated what was perceived as the beginning of an extensive crisis in computer science, mainly due to a shortage of manpower, especially for the low number of PhD students, despite an exceptional increase in undergraduate enrolments. In the early 1980s, recent graduates were looking for positions outside universities, with a reduced number of academic positions available and an increasing pressure from industry to acquire them. The report established at around 60% the portion of CS graduates receiving a job in industry, attracted especially by high salaries. In this context, the report presents computer science as

both a theoretical and an experimental science. In this it is similar to the physical sciences. Computer Science is also an indispensable tool in other disciplines. In this it is like the mathematical sciences.[5]

---

[2] See (Feldman and Sutherland, 1979, p.498).

[3] (Feldman and Sutherland, 1979, p.497).

[4] Denning et al. (1981).

[5] (Denning et al., 1981, p.372).

The main aim of the report was to make the case, especially to the government, for a significant increase in funding in order to guarantee the necessary influx of high-quality graduates in the PhD programmes, and to preserve and strengthen the educational capacity in the engineering and computing fields. But it is notable that, while computing was associated with some of its experimental aspects earlier, the comparison is now directly made with the industry: at this moment in time, computing is recognized as a discipline that has a strength and ability of production outside the purely intellectual and scientific fields, and that companies have by now established a market which also creates educational needs and skills requirements that cannot be determined purely as a function of knowledge.

The notion of experiment in CS thus started to appear, but often used with different meanings by different people. In the above cases, it was used to refer to both the practical and industrial use of computational technology, but also to its ability to provide support to other sciences in terms of computational techniques. The terminology on experiments simply followed the use of other sciences, but with little attention to the specific aspects of the discipline and its methodological complexity. Notoriously, a crucial aspect of the experimental methodology of the hard sciences is the hypothetical-deductive method: the positioning of hypotheses and their testing under controlled conditions. This aspect was not overlooked for too long in computing either. Soon the methodology of computing was linked with simulation and prototyping techniques,[6] two aspects that will receive increasing relevance in later decades. The role and importance of experimental computer science were linked to four factors:

1. the growth of systems, both in size and speed;
2. the exponential growth and invasiveness of computing technologies;
3. the increasingly facilitated interaction with the user, despite lack of expertise and expectation;
4. and finally the strengthening of the methodological aspects of the discipline, transforming it from an engineering art to a proper science.

This last factor was essentially due to the pursuing of its experimental nature, crucial in reducing malfunctioning and increasing reliability. For the proponents of CS as an experimental science, the benefits were clear. On the one hand experimentation allowed current systems to be improved on, especially in those settings like real-time and distributed systems where the help of theoretical research, although foundational, was not sufficient. On the other hand, experiments also helped improve the scientific image of the field, and this had an impact on increasing funding, thus reducing the associated problem in academic development.

There were also strong critiques on the experimental aspects of CS. In particular, it was becoming obvious that the nature of computational experiments was unclear, in which

---

[6] See e.g. Schorr (1984).

way they differed from experiments in the natural sciences and how the scientific method required to be adapted for the specific aspects related to computing. Nor was the role of experiments detailed in understanding the nature of the discipline. Moreover, the view of computing as a mathematical science requiring an axiomatic view was still strong, as well as the view of computing as an engineering art requiring physical principles, and these both often clashed with its experimental interpretation.

Still, some decades after these early attempts at characterizing the experimental nature of computer science, the analogy between the scientific method and the problem-solving process underlying computing still is a tempting proposition. The hypothetical-deductive method is rooted in the philosophical doctrine of empiricism, formulated in the eighteenth century by John Locke (1632–1704). The school maintained that all knowledge comes *a posteriori* from experience and justification, according to which observations lead to hypothesis formulation, in turn allowing theory construction about reality if appropriately tested and verified. Confirmation of hypotheses has been underpinned by different philosophical positions. Verificationism (or logical positivism) was based on the empiricism of George Berkeley and David Hume and developed by the philosophers of the Vienna Circle and in particular by Alfred J. Ayer and Rudolf Carnap. Verificationism argued that empirical hypotheses do not require complete verifiability, but that experience is possible to attest their truth or falsity, thereby excluding metaphysical propositions. Falsificationism was formulated by Karl Popper with the aim of arguing that knowledge does not require evidence to prove a theory to be true but rather evidence that would prove a theory to be false, through what he called *potential falsifiers*. From these two main approaches (and others, like the paradigm model), the modern concept of scientific discovery is formulated. The hypothesis and testing method aims at making discoveries about the empirical world, through the formulation of hypotheses that can be tested against regular patterns of observable events.

It is beyond the scope of this chapter to reconstruct the details of the large debate on the hypothetical-deductive method and its strenghts and weaknesses. Our task is to attempt a clarification of the characteristics of hypotheses and experiments in the context of the computational sciences. The debate on the difficulty of designing hypotheses and performing appropriate experiments for computer science is not new.[7] The first hurdle on the use of experiments in computing, in particular related to proving properties like correctness and reliability, is their limited ability to act as verifiers, and the strong falsificationist value they bring. But, undoubtedly, experiments are useful in the context where proofs are of limited applicability. Since the early days of computing, and explicitly so already in the Feldman Report, experimentation in computing has included exploration, hypothesis construction and testing, demonstration, and modeling. Nonetheless, the definition of the notions of hypothesis and experiment specific for the computational sciences have not been formulated yet. One comparative approach to the definition of

---

[7] For a general introduction to the debate on the hypothetical-deductive method, see https://plato.stanford.edu/entries/scientific-method. For a review of the design of experiments in computer science, see Tedre and Moisseinen (2013).

computing as an experimental discipline consists in presenting it as a combination of both engineering and an exact science:[8]

> [I]n some circumstances computer science is best viewed as the engineering, possibly using mathematical methods, of problem solutions on a computer, while in other circumstances it is best viewed as concerned with the experimental testing of hypotheses and the subsequent recording of observations in the real world.

This view relies on the incompleteness of program testing for complex systems, where a rigorous methodology intervenes to maximize test case coverage, or to justify a mathematical approach. A purely experimental approach requires insistence on more careful testing practice, where the inability to prove the absence of errors (Dijkstra docet) is not a sufficient reason to reject the validity of a system. Hence it does not weaken the validity of the methodology either. The possibility of finding a defeating instance of a test suite is in fact embedded in the experimental approach, and in the case of programming this is characterized by attempting at individuating typical test inputs. This reduces, in most real cases, to the fact that abstract machines allow typicality assumptions far easier than programs executed on physical machines. Additionally, it requires an assumption that hardware and software on which the tested program is running behave correctly, i.e. the system satisfies physical correctness as defined in the previous section.[9]

In this context, the scientific method seems to offer a parallel with the algorithmic method in computer science: the scientific method consists in formulating a hypothesis for explaining a phenomenon, testing the hypothesis by conducting an experiment, and finally in confirming or rejecting the hypothesis by evaluating the results of the experiment; analogously, problem-solving in computing can be seen as formulating an algorithm for solving a problem, testing the algorithm by writing and running a program, and finally in accepting or rejecting the algorithm by evaluating the results of running the program.[10] In this way, testing is an experimental process aimed at determining whether an algorithm solves a given problem. The analogy fails in the nature of programs, which are more and more conceived as black boxes and therefore less prone to the intervention by the experimenter.[11] Under this view, problem-solving in computing can rather be seen as formulating a solution design, in terms of a program specification, for solving a problem, testing the solution design by generating and running a program, and finally accepting or rejecting the solution design by evaluating the results of running the program. This interpretation has the advantage of admitting both testing and simulation as viable experimental methods for the discipline.

Hence, beyond the practical aspects of the different methodological approaches to the development of software systems (Waterfall, Spiral, Agile and so on),[12] the properties

---

[8] (Colburn, 2000, p.154).
[9] See also (Colburn, 2000, pp.160–3).
[10] See (Colburn, 2000, p.168).
[11] This is observed in (Colburn, 2000, pp.169–70).
[12] See e.g. (Turner, 2018, Ch.13).

of the notions of hypothesis and experiment required by a methodological reading of computing still need to be qualified in detail. This is the task of the following sections of this chapter.[13]

13.2 On Computational Hypotheses

Computational experimentation usually refers to a formulation of standard scientific hypotheses where the system under experimentation is virtual, or simulated, or expressed through computable equations. The term itself is not unusual in fields like neuro-biology and neuro-physiology, but it can be used to indicate any claim formulated and then tested in computer science research, concerning for example the invention of a new technique or algorithm, the improvement or extension of an existing technique, a new application of it, or a better understanding of its properties. Usually, such claims are tested behaviourally, i.e. by analysing how the system or systems under scrutiny behave when operational on some input of interest and in a controlled environment. Alternatively, computational hypotheses in this sense are tested in relation to the extent to which they apply, i.e. in reference to the class of systems satisfying them, with part of this experimental analysis aiming at establishing whether the hypothesis holds for systems beyond those assumed to belong to the class under scrutiny. Finally, they can be tested with respect to how efficient those systems need to be.

Along with this uncontroversial way of referring to hypotheses in a computational setting, a stronger epistemic sense in which this notion can be understood is by reconsidering standard scientific principles of hypothesis formulation, experiment design, and experiment execution. Reformulating these notions for the context of computing as an experimental science means providing an interpretation of them as computational statements. For the case of hypotheses, we shall call these statements *computational hypotheses*. After the recognition of a situation that suggests the formulation of a problem, in empirical research hypotheses are typically formulated according to the following types:

1. Declarative hypothesis: a positive statement expressing a relationship between the variables of interest;

2. Null hypothesis: a negative statement expressing that no significant difference exists between the variables of interest;

3. Alternative hypothesis: a hypothesis contrary to the null hypothesis;

4. Interrogative hypothesis: a question about certain relationships between the variables of interest;

5. Predictive hypothesis: a positive statement expressing expected principles emerging from testing.

---

[13] Our proposed analysis aims at remaining neutral with respect to the practical development method used, but especially as to whether the underlying methodology of any such process should be deductive or abductive.

Our attempt is therefore to provide counterparts to these notions which are explicitly formulated for the context of computing. Let us start from the first meaning. We shall provide two versions:

Definition 130 (Computational Declarative Hypothesis — Black Box Version)
A statement about the existence of an algorithmic behaviour expressing a given I/O relation.

This hypothesis is characterized only as a statement that an algorithm exists to the effect that given a certain input, a certain output can be obtained. This formally corresponds to the decision problem about the truth of a given statement, where the statement corresponds to the specification of the algorithmic system under consideration. As such, the Black Box Version qualifies as a semi-decidable problem. If the Black Box has been shown to be positively answered, the White Box Version of the Computational Declarative Hypothesis can be formulated:

Definition 131 (Computational Declarative Hypothesis — White Box Version)
A statement about the structure of an algorithmic behaviour expressing a given I/O relation.

As said, the White Box Version of a Computational Declarative Hypothesis is conditional on the corresponding Black Box Version being answered positively. Then this version is decidable, because given any existing algorithm whose I/O behaviour is known, it can be checked whether it corresponds to the structure stated in a White Box Computational Declarative Hypothesis. There is, however, a different dynamics illustrated by testing: the design of an algorithm which answers a White Box Computational Declarative Hypothesis can precede the formulation of a corresponding Black Box Version, which then gets answered by the existence of an algorithm.

If a Computational Declarative Hypothesis cannot be answered, at some point a Computational Null Hypothesis (to an equivalent I/O relation) can be formulated:

Definition 132 (Computational Null Hypothesis) *A provable statement that a certain I/O relation does not hold.*

This means hypothesizing that no algorithm can be defined such that given a certain input a corresponding output can be produced. This is obviously the negative counterpart of a Black Box Computational Declarative Hypothesis and as such equally semi-decidable: i.e. it can only be answered in the positive if a counterexample is provided such that the computational algorithm in question is shown to be impossible. This corresponds to the hypothesis that no algorithm exists such that given a procedure as input, it returns a yes answer if that procedure qualifies as an algorithm for a Turing Machine, and it is answered negatively otherwise.

The negative counterpart of the Computational Null Hypothesis, i.e. one that is formulated if the Null Hypothesis is answered negatively, is (not surprisingly, given the semi-decidable nature of the problem) still an hypothesis:

Definition 133 (Computational Alternative Hypothesis) *A provable statement that a certain I/O relation holds.*

Here the hypothetical process works as follows: assume it has not been possible to disprove the existence of a certain I/O relation; is it possible *to prove* that such a relation holds? Note that the stress on provability renders *possible* tests proving the hypothesis insufficient from a methodological point of view: what is required is, instead, absolute certainty, i.e. proof that the relation holds.

The answer to the Computational Alternative Hypothesis reduces, essentially, to the computational variant of the Interrogative Hypothesis:

Definition 134 (Computational Interrogative Hypothesis) *Does a certain algorithmic relationship hold between the input and output variables of the problem of interest?*

This formulation presents two different aspects for clarification. First, once the input and output variables for the problem of interest have been fixed, then it is essential to qualify which type of algorithmic relation holds between them (rather than just any given relation, as long as it can be algorithmically defined): in other words, the question concerns here a certain algorithm with efficiency and optimality (complexity) qualities. Second, when both variables and relation of interest have been fixed, the Computational Interrogative Hypothesis is essentially equivalent to the Type-Checking Problem (see Definition 101) where the derivability relation matches language and complexity properties of the algorithm of interest.

Finally, the following is possible:

Definition 135 (Computational Predictive Hypothesis) *A prediction about how a system will behave algorithmically (including output expectation) when provided with a given input.*

The first strong version of this hypothesis depends on no previous hypothesis having been answered. In this case, the only variable of interest known is the input; both structure and properties of the algorithm are not fixed; then one can anticipate (depending on the amount of background knowledge on input and system, this ranges from speculation to prediction) on both output and properties. More realistic is a weaker version: assume the existence of a certain I/O relation has been established through a positive answer to both Black and White Box Versions of the Computational Declarative Hypothesis; assume moreover that a version of the Computational Interrogative Hypothesis has been formulated, albeit not yet necessarily answered; then a Computational Predictive Hypothesis, while not concerning the output expectation any longer (as that is supposed to be answered by assumption), refers instead to the behaviour of the algorithm, for example in which complexity class will it be located.

The computational interpretation of hypothesis confirmation through the verification of its consequences follows. For a computational hypothesis, we need to identify its consequences as algorithmic behaviours implied by a given assumed I/O relation. Obviously, a behavioural closure of a given system on a given output can be assumed but verified only up to a given level of certainty. Epistemically stronger is therefore the notion of a computational falsifier, corresponding to an observed behaviour which rejects a given hypothesis. In this case, we consider two appropriate notions:

Definition 136 (Black Box Falsifier) *An instance of an I/O relation which, given an input appropriate to a computational hypothesis, returns an output different from the one expected.*

Definition 137 (White Box Falsifier) *An instance of an algorithm different from the one defined according to the computational hypothesis, which realizes the intended I/O relation.*

Note that the Black Box Falsifier cannot be assumed to reject the corresponding Black Box Computational Declarative Hypothesis, but only that algorithmic behaviours exist which are non-deterministic on the given input. Similarly, its dual (that an instance of an algorithmic behaviour different than the one formulated by the White Box Computational Declarative Hypothesis produces the assumed I/O relation) only offers an extension of the hypothesis on the possible algorithmic behaviour required to satisfy the intended relation.

A further clarification of the nature of computational hypotheses comes from a comparison with properties of experimental hypotheses proper of the scientific method analysed in the philosophy of science. These often go back to the debate between verificationists and falsificationists in the 1960s and later. One such property was proposed by Nelson Goodman, arguing that any set of observations will equally support multiple hypotheses.[14] In our formulation, this concerns the abstraction of a Black Box Declarative Computational hypothesis from a I/O relation:

Thesis 11 (Equivalence from Observation) *A set of observed I/O relations will equally support multiple hypotheses on the underlying algorithmic behaviour.*

This thesis needs to be understood modulo the preservation of logically valid operations on the signature of the I/O function and it is not enough that the multiple admissible hypotheses preserve an homomorphism on the I/O pairs, rather it is necessary that the two inputs and the two outputs be logically equivalent in their types.

This property was further clarified by Carl Hempel, who argued that an observation confirming a given hypothesis also confirms all its logical equivalences.[15] In our computational translation, this thesis is related to the formulation of a White Box Declarative Computational Hypothesis from a I/O relation:

Thesis 12 (Logical Equivalence from Observation) *An observed algorithmic behaviour confirming a hypothesis about an underlying I/O relation, also confirms all the logically equivalent behaviours.*

This thesis requires reference to a transparent understanding of the structure of the algorithm (White Box) because behavioural equivalence cannot be understood at the level of function signature if it has to preserve step-wise logical identity. Also note that the formulation of a White Box confirmation is much stronger than the confirmation

---

[14] Goodman (1965).
[15] Hempel (1965).

of a test implication, which according to Hempel could only provide corroboration or confirmation for the hypothesis, but it is not sufficient proof that it actually holds.[16]

This induces a further property, which Popper formulated in terms of the informational content of a hypothesis as directly proportional to the amount of its falsifiers, i.e. the more realistic the hypothesis, the more the falsifiers needed to reject it.[17] The relation between hypothesis and its falsifiers can be strengthened for their computational counterparts:

Thesis 13 (Informational Content) *The informational content of a White Box (resp. Black Box) computational hypothesis is inversely (resp. directly) proportional to the amount of its Black Box (resp. White Box) falsifiers.*

Given a hypothesis about the structure of an algorithm satisfying a given I/O relation, i.e. a White Box declarative statement, a significant amount of Black Box algorithms falsifying that relation is required to weaken the value of the hypothesis. Hence, in line with Popper's view, the more potential falsifiers of a hypothesis, the more falsifiable it would be, and more would be claimed by the hypothesis if it stands trial. Adapting another Popperian intuition to the case of refutation: despite the amount of refuting evidence, we can never be certain that a computational hypothesis does not hold if our only evidence to the contrary are Black Box falsifiers. In this case, we can talk of *falsifying corroboration* as a measure for how weak a hypothesis has become after negative testing — although this cannot imply a measure for the probability that it is false, unless a proof of the impossibility of the related White Box declarative statement can be offered. Conversely: given a hypothesis about the existence of an algorithm satisfying a given I/O relation, a single transparent algorithm that falsifies that relation is sufficient to nullify the value of the hypothesis: in other words, a Black Box Declarative Hypothesis requires only one White Box falsifier to reject it. In this case, the less the potential falsifiers required to reject the hypothesis, the more the falsifiers themselves will have to claim in order to be accepted.

The next step in our analysis is a clarification of the experimental method applied following the formulation of a computational hypothesis.

13.3 On Computational Experiments

In the scientific method, once the hypothesis is formulated, an appropriate research method needs to be established (statistical or experimental) and testing has to be deployed to confirm or disconfirm it. Similarly, once a computational hypothesis is formulated, testing and simulation can be deployed as appropriate methods to corroborate or reject it. In experiments, all variables manipulated and controlled are called independent variables; the variables studied to see the effect of the changes in the independent variables are called dependent variables. Our interest is now in comparing properties of the physical experiment apt for testing and simulation in a computational setting.

---

[16] (Hempel, 1966, p.8).
[17] Popper (1959).

The observation that computing—and more specifically software engineering—was in need of the definition and integration of components of an appropriate experimental process was made early in the history of the discipline. A detailed analysis of the experimental process was offered already in the 1980s.[18] According to this view, the process starts with a study definition phase (problem, objective, goals), followed by a study planning phase (instruments selection, threats consideration), a study operation phase (measurements collection), with a final phase of interpretation (measurements analysis).[19] This intuition was further developed in the following decades. The components of the experimental process have been summarized in terms of four parameters:[20]

1. object of study: what is the system for which a model is sought;

2. purpose: for which aim is the model designed

   - definition,

   - evaluation,

   - prediction,

   - control,

   - improvement;

3. focus: the aspect of interest of the system

   - reliability,

   - defect prevention,

   - accuracy;

4. point of view: the user benefiting from the analysis.

In this context, experiments are analysed as descriptive (searching for patterns in data), correlational (searching for relations between independent and dependent variables), cause-effective (searching for cause–effect relations between independent and dependent data).

The different types of experimentation in computational settings from the current literature can be summarized as follows:[21]

   - Demonstration: showing that a system is feasible, reliable, cost-efficient, through construction and demonstration;

---

[18] See Basili et al. (1986).

[19] See also (Wohlin et al., 2000, sec.2.4.2 and sec.4.2), where results presentation and packaging are added. In Denning (1981) performance evaluation was illustrated as an excellent form of experimentation in computer science, referring to two study cases, namely time-sharing in the M44/44X project at IBM Watson Research Lab in the mid-1960s and queuing models until the mid-1970s.

[20] See Basili (1996).

[21] For this summary, see especially (Tedre and Moisseinen, 2013, sec.3); for the explorative meaning of experiments, see Schiaffonati (2016).

- Trial: showing that a system meets some specification through a test (in situ, emulation, benchmarking, simulation);
- Testing: showing a functional system in a live environment (in situ, or in vivo);[22]
- Comparison: showing that a system S outperforms another system S' on a given task, with a given dataset and parameters;
- Control: standard application of the scientific method, to show that a given computational hypothesis holds;
- Exploration: experimentation is not theory-driven, but explorative in the sense of a process of testing technologies in their socio-technical contexts, with control applied a posteriori.

The evaluation of computational experiments has been widely debated. The first definitional step in the experimental process is given by (independent and dependent) variables selection, with corresponding measurement scale, range, and tests.[23] A representative selection of examples is a standard issue of experimental science, and it converts to experimental computing science as well by translating to algorithms and their I/O relations. The standard solution should be adapted by separating a training from a test set of inputs of the required type for a given algorithm: to test whether the algorithm actually satisfies the intended I/O relation, a varied, unbiased, generally acknowledged, and well-designed set of input examples must be chosen, to check it always returns the intended output. The selection can be done by random, systematic, stratified, convenience, and quota sampling. The validity of such a set of examples is determined by evaluating them through White or Black Box systems, i.e. by the experimenter's ability to test inputs on systems qualified by their internal functional mechanisms. Size of the sample should be considered, with wide variability in the population requiring a larger sample size and the analysis of the data possibly influencing the choice of the sample size.[24] Controlled experiments on the basis of such an example set should be compared with sufficiently similar algorithmic systems for differences. Similarity needs to be defined in terms of knowledge of the algorithmic structure, languages used and physical support deployed. Each of these three levels of abstraction (algorithm, language, physical system) has to be taken into account and the comparison is meant to reduce the inherent difficulty of controlling systems. In this, the design of common methodologies of experimentation should be assumed.

The selection on an example set of inputs for an algorithm leaves open the question of which experimental method should be applied. In the present context we consider two cases. The first methodological approach is that of proper testing. Software testing is normally considered as the practical investigation on the product quality and properties, and we have considered it to some extent in Chapter 10. In the present context we are interested in which properties testing requires and induces as a form of experimentation.

---

[22] This was also called *observational experiment* in Basili (1996).

[23] See (Wohlin et al., 2000, sec.6.3.1–2).

[24] See (Wohlin et al., 2000, p.50).

The first observation is that Computational Declarative Hypothesis types have each a corresponding testing method (i.e. Black or White Box): with the former, one is interested in testing functionalities, with the latter to test mechanisms. Some approaches have stressed certain commonalities of testing with science: the reliance on falsifiability to prove experimentally that software works, the skeptical approach of testers, the underlying notion of knowledge of a system as a systematically weak one, the combination of skills, inferential and conjectural methods following the design of a model of the system.[25] In general, though, testing consists of guiding principles and it is not a strict method. Accordingly with the strong use of falsificationist procedures, the above notions of White and Black Box Falsifier and the above analysis of a varied example set especially apply to testing in order to ensure some level of reliability. The most important observation, though, is that testing requires essentially a software and hardware artefact construction, and as such it crucially involves property generalization.

The second case of computational experimentation is represented by computer simulations. The use of simulations as experiments minimizes the reduction of the experimental process to an analysis of cases. Instead, it increases the inferential nature of the process, mainly based on the formulation of a model for the target system. Nonetheless, simulation still requires the implementation in a computational model, i.e. the translation in a language and its execution on hardware, which will induce the same limitations of any other computational artefact in terms of the required generalization. For this reason, the main problem in establishing a simulation model sufficiently apt for extrapolation of valid results is to provide a balanced design between generality and precision.[26]

Assuming either testing with sufficient coverage or simulation with a balanced model, the validity of the experimental results extracted from each can be assessed either with respect to the sample input population only, or generalized to a broader input population, on the basis of the analysis of systems compared for similarity. A taxonomy of validity principles can be formulated for computational systems as follows:[27]

Principle 14 (Conclusion Validity) *Validity concerned with the relationship between the experimental method chosen (testing or simulation) and its output.*

Principle 15 (Internal Validity) *Validity concerned with the relationship between the causal relationship between method chosen and output.*

Principle 16 (Construct Validity) *Validity concerned with the relation between model and observed behaviour in the chosen method.*

Principle 17 (External Validity) *Validity concerned with the generalization of results from the observed behaviour to the model.*

---

[25] For an analysis covering these aspects, see e.g. Kaner et al. (2001).

[26] The relations between target system, formal and computational model, and artefact implementation as simulation will be explored at length in the next chapters.

[27] This list is inspired by the one offered in (Wohlin et al., 2000, pp.64–5), and based on Cook and Campbell (1979).

The construction of models and their implementations remain the most important hurdles in the design of computational experiments, their validation and evaluation. In Chapter 14 we will further investigate these essential problems with particular attention to the case of computer simulations.

Exercises

Exercise 122 *Reconstruct the meanings of experimental computer science indicated in the Feldman and the Snowbird Reports.*

Exercise 123 *Which critiques can be applied to the understanding of computing as an experimental discipline?*

Exercise 124 *How can computing be connected to the experimental method known from the traditional sciences?*

Exercise 125 *Explain the difference between Black and White Box Computational Declarative Hypotheses and illustrate their connection with appropriate falsifiers.*

Exercise 126 *In which way are the Computational Null Hypothesis and its negative counterpart related to decidability properties?*

Exercise 127 *In which sense is the Computational Interrogative Hypothesis resulting from a positive answer to the Computational Alternative Hypothesis, and why are complexity properties necessarily related to it?*

Exercise 128 *In which versions can the Computational Predictive Hypothesis be formulated?*

Exercise 129 *Explain the computational version of Goodman's thesis that any set of observations will equally support multiple hypotheses.*

Exercise 130 *How is observational equivalence related to logical equivalence in algorithmic behaviours?*

Exercise 131 *Relate logical equivalence of algorithmic behaviours to their informational contents.*

Exercise 132 *Illustrate the different types of experimentation in computing.*

Exercise 133 *Clarify the required properties of controlled computational experiments.*

Exercise 134 *Explain validity of controlled computational experiments.*

14 Models and Simulations

Summary

In this chapter, we further investigate the constitutive elements of the experimental foundation of computing: we first introduce the different notions of models, then overview the types of computer simulations from the literature, and finally explore the various purposes attached to their use as implementations of computational models.

14.1 On Models

Understanding experimental computing requires an illustration of the debate on modelling for scientific purposes. Experimental science makes extensive use of modelling practice and of the process of abstraction from it: in particular, the construction of a model and the execution of controlled experiments to test its validity are an obvious, essential, and methodologically sound choice to explore a complex system when its empirical construction is non-trivial, too expensive, or dangerous. Different kinds of modelling types can be identified to this aim:[1]

- *Concrete Models*: these are physical, constructed or naturally occurring, concrete scale models which stand in a representational relation with their intended targets; examples are: the San Francisco Bay model; the use of model organisms like fruit flies in biology; natural experiments in population dynamics, geology, climatology, and anthropology;[2]

---

[1] An extensive overview of these kinds of models and their construction is offered in Weisberg (2013).
[2] See (Weisberg, 2013, ch.1.1 and ch.3.1.1).

On the Foundations of Computing. Giuseppe Primiero, Oxford University Press (2020). © Giuseppe Primiero.
DOI: 10.1093/oso/9780198835646.001.0001

- *Mathematical Models*: set-theoretic predicates or state transition systems associated with dynamical equations completely describing the properties of the target system; examples are thermodynamic, classical, or quantum mechanical state spaces, determining variable quantities in algebraic expressions possibly with fixed parameters and temporal dimensions;[3]
- *Computational Models*: algorithmic structures, procedurally defined; their most important properties are agent-based determination of the procedures and conditional structure, possibly enhanced by probabilistic and parallel transitions; the computer-based implementation of these models is an additional but not strictly essential feature.[4]

In the context of our investigation on the experimental foundation of computing, we consider in particular the relation between mathematical and computational models and the implementation of the underlying algorithmic structures in programs. The latter will be mainly identified with simulations.

The variety of modelling practices implies also a variety of study conditions and aims. From a methodological viewpoint, the conditions of the modelling exercise change in relation to the target system. In target-directed modelling, the starting assumption is that a model is compared to a real-world target. To this aim, there are essential steps to perform.[5]

1. *Model Development*: whatever the type of model at hand, the selection of the appropriate structure, the determination of its expressiveness and its interpretation, including its evaluation, are required. Structure selection is dictated by the required properties of the intended model and target; expressiveness evaluation is necessary for the modelling aim and in particular to guarantee explanation or prediction; the interpretation in terms of formal or informal properties determines the degree of generality and applicability of the model, both in terms of different systems within the same family, and across largely different disciplines, as well as the limits of the admissible deviations. Note that the interpretation of the model can be a dynamic process: if the target is pre-determined, this process must converge towards the representation of such system for the modelling process to be judged successful; in contexts where the modelling practice is intended to help building the target system, e.g. in robotics, the dynamic construction of the interpretation can proceed further, with the only constraints to the model being determined by the environment in which the intended system to be constructed has to exist.

2. *Model Analysis*: it corresponds, roughly, to determine (analytically, by numerical analysis or experimentally through simulation) the behaviour of the target system under a full description of the model (i.e. of its states and transitions). Note that model evaluation can be partially dependent on the method used and, in particular,

---

[3] See (Weisberg, 2013, ch.3.1.2).
[4] See (Weisberg, 2013, ch.3.1.3).
[5] For an introduction to these steps, see (Weisberg, 2013, ch.5).

if the numerical analysis starts from specific initial conditions of the system under consideration.

3. *Target Calibration*: required for a model designed with a specific goal, or for the adaptation of a generic model to a concrete domain or problem. This is usually obtained by the maximization of an identified objective or fitness function, through the identification of relevant variables and parameters for the starting conditions and the search of the related space (by full exploration, random search, hill-climbing techniques, genetic algorithms, and so on). Whenever the instantiation of a model is not directly feasible, approximation techniques through iterative formulation of less and less idealized versions of the model can be put in place, or by perturbations that add or remove components.

Our interest in the process of model development, analysis, and target calibration is dictated by two possible aspects: first, the process of design and analysis is aimed at a computational model (and in particular an implementation thereof); second, the model designed and analysed is targeted for a computationally characterized science or application, e.g. network theory, or robotics. The methodological advantages offered by the construction of mathematical models, their translation to computational models, and the implementation of the latter in simulations for experimentation consist in the explanatory ability of the modelling process, whose qualification depends on the type of formal underlying relations. Computational experimentation further facilitates the exploratory aim of the model, in particular for complex systems that cannot be represented in other types of models.

In the following, we will be interested in the relation that formal models have with their target systems, and how computational models relate to their implementation. As the above summary illustrates, in the literature on scientific modelling the main aim of model building is usually identified with the ability of models to provide explanations of their target systems. This makes models an essential element in the philosophical debate on the explanations of laws, with a variety of theoretical stands. Some pragmatic understandings of this relation disregard the role of explanation in assessing the value of a theory and in turn assume that laws do not require explanation.[6] Causal interpretations of models assign to them descriptive power over the causal relations between facts and processes, and identify these relations as having explanatory power.[7] Accounts that attribute directly to models an explanatory power require them to fit into the basic framework of a theory;[8] a different way of characterizing the explanatory power of models is by correlating better understanding with better unification.[9] Models can also be qualified as (semi-)autonomous mediators between abstract theory and concrete systems, such that

---

[6] For this position, see e.g. Fraassen (1980).
[7] See Woodward (2003).
[8] See e.g. Cartwright (1983).
[9] See Kitcher (1989).

the former is capable of yielding information about the latter.[10] The role of models in providing explanations about their target systems can be defined as the task of finding the right balance to mediate between the idealization provided by the mathematical formalization and the direct, empirical knowledge of the system under investigation. This approach needs to be qualified in order to understand the possible implications of different levels of mediation and which relations hold between formalization and empirical constructions.

A stronger sense in which models relate to systems consists instead in manifesting predictive power, intended as the prediction of new or future observations about the target system. In the hypothetical-deductive method, explanation and prediction are usually equated, as the latter can be defined by the deductive closure of the statements fully explaining the system of interest. On the other hand, in statistical methods the uncertainty associated with prediction is different from what is usually associated with explanation, as the latter typically assumes a temporal characterization distinguishing data and the phenomena they represent: the formulation of models creates a separation between explaining phenomena and the ability to make future predictions at the measurable level.[11] Typically, a major problem of predictive models has been the identification of their accuracy.

From our point of view, the distinction between explanation and prediction reduces to the definition of relations between formal and computational models and their relation with the target system (i.e. the phenomena): on the one hand, the formulation of models (formal or algorithmic) that are simply isomorphic to the target system expresses an explanatory ability about the phenomena; on the other hand, the formulation of models for which also every admissible computational behaviour is guaranteed to find a counterpart in the phenomena, will express precisely the idea of predictive power of the model as its ability to express a full specification of the target system, together with inferential abilities over it.

A different type of modelling activity is involved in the design of a model which does not aim at the full specification of a predetermined target system but rather matches one of the following two cases:

1. either the designer aims at identifying a model (within a given class) that formulates the best approximation to the target system; in this case, there is no intention of mimicking precisely the behaviour of a system but rather to repurpose certain features known from a given existing model to a different context: a typical example is the design of swarm robotics systems inspired by and mimicking features of natural models, e.g. of bees or ants;

2. or the designer aims at constructing a model approximating a given (possibly ideal) system, in the sense of including a behaviour or property common to a predetermined system, while at the same time being compatible with a number of

---

[10] See Morrison (2015).
[11] See e.g. (Shmueli, 2010, p.293).

extensions; in this case, there is a task requiring a certain fixed feature which can be explored in combination with other tasks, like the design of protocols in network theory where the underlying mathematical structure of a graph is combined with behavioural properties of agents and the aim is to explore the configuration that maximizes certain features (e.g. the distribution of information).

Both cases can be identified as forms of modelling without a specific target system.[12] In these cases, approximation can be based either on a non-Bayesian simplicity criterion for information, or on prior probability distributions (i.e. a Bayesian account), with constraints like language invariance (across the class of models of interest), independence from cost functions, generality, consistency, and efficiency.[13] But approximation is also appropriate to refer to an exploratory behaviour of certain modelling activities in the context of the artificial sciences, like robotics, where the development of a system can be based on the exploration of possible combinations of subsystems guided by simulation activities. Our analysis, focusing on the relation of models with systems as a way to define experimental computing, is restricted to the latter case: in particular, we will concentrate on the relation between computational models and their implementations and will use simulations as (non-exclusive) representatives of the class of artefacts instantiating models. In Section 14.2 we overview the various types of simulations.

14.2 On Computer Simulations

The practice of computer simulation has received an increasing amount of attention in the recent philosophical literature. A first distinction needs to be drawn concerning the scope of this term. A first investigation concerning the epistemic novelty represented by simulation with respect to experiments refers to a broad sense of simulation:

> In the broad sense, 'simulation' refers to the entire process of constructing, using, and justifying a model that involves analytically intractable mathematics [...] Following (Humphreys, 2004, pp. 102–104), we call such a model a 'computational model'.[14]

A simulation in this sense is meant not only as the artefact which implements a model, but it indicates the whole process of experimentation. This broad sense referring to a comprehensive method and process for studying systems, includes:

- designing a formal model;
- translating that model to an algorithm;
- implementing the algorithm in a language;

---

[12] See (Weisberg, 2013, ch.7).
[13] For these criteria, especially with respect to compact coding, see Wallace (2005).
[14] Frigg and Reiss (2009).

- devising an experiment based on running the algorithm on a computer;
- calculating the output of the algorithm;
- and studying the resultant data (possibly through visualization);

and it corresponds precisely to the experimental framework of computing which we are exploring and defining in this part of the volume.

On the other hand, a narrower sense of a computer simulation refers to a program that is run on a computer and that uses a step-by-step method to explore the approximate behaviour of a mathematical model.[15] The narrow sense of simulation must be understood in terms of the linguistic and physical implementations of an algorithmic model, with the proviso of validity for physical computation explored in Part II of this volume. The algorithm must be an appropriate translation of an underlying formal model of the (possibly hypothetical) target system. The algorithm's input is a specification of the system's state at some initial time t and the simulation aims at calculating the system's next state. The output is obviously the state of the system at the end of the transformations of interest. In this sense, the approximation from continuous to discrete solutions results also from the specific implementation and execution, with different results possible under different implementations and physical realizations. This more strict and precise sense of the notion of simulation can be formulated as follows:[16]

Definition 138 (Core Simulation, Humphreys (2004)) *A system S provides a core simulation of an object or process B just in case S is a concrete computational device that produces, via a temporal process, solutions to a computational model [...] that correctly represents B, either dynamically or statically. If in addition the computational model used by S correctly represents the structure of the real system R, then S provides a core simulation of system R with respect to B.*

A successful simulation allows therefore for experimenting on an algorithmic model of a physical system (ideal or real) to investigate the behaviour of its components. The success of this experimental practice is a function of our knowledge of the parameters involved in the modelling of the system and its complexity, and in the strict sense of the term also of the properties of the linguistic encoding available and of the physical device in use.

In their narrow definition, computer simulations are characterized by their methodology:[17]

- *Particle-methods simulations*: particle methods reflect a type of simulation allowing both discrete and continuous systems. Particles are abstract structures that can represent both agents and mathematical discretization points for solving differential equations. Their versatility allows treatment of different types of models by the same algorithms and data structures. The granularity of particles can vary to include superparticles representing large number of the real elements of interest, or instead

---

[15] For these definitions of simulation in the broad and narrow sense, see (Winsberg, 2015, sec.1.2).

[16] For this definition, see (Humphreys, 2004, p.110).

[17] For this list, see Winsberg (2015); Woolfson and Pert (1999).

the simulation can estimate the overall field of each particle in relation to others through the definition of cells, geometrical regions or volumes of materials as periodic structures all with the same contents. Particles can be defined by quantities associated with the matter of interest and the change of forces applied to them.

- *Finite-difference methods simulations*: devised to investigate properties through the definition of a region of interest and by the approximation of differential values by linear combinations of time-based grid-point values to be determined.

- *Finite-element methods simulations*: in this case the definition of a region of interest is obtained by points (nodes) and their connections defining elements, and multi-dimensional shapes with boundaries; the method aims at finding the values for the quantity of interest at the nodes in an equilibrium.

- *Agent-based simulations*: most common in the social and behavioural sciences, but also largely used in artificial life, epidemiology, ecology, robotics, and any discipline relying on a network of individuals in interaction. Although close to particle-based simulations for their studying *n*-many discrete individuals (agents), they differ because global differential equations that govern the motions of the individuals are missing. Instead, in agent-based simulations, the behaviour of the individuals is dictated by their own local rules and their interaction with the environment. The equations are then used to describe a dynamic or evolutionary process, with the specificity of the characteristics of the space in which the interaction occurs. Agents are usually characterized by:

  - *Autonomy*: agents are autonomous information processing and exchanging units, free to interact with other agents;

  - *Heterogeneity*: agents may have different properties and be grouped according to similar characteristics;

  - *Activity*: agents are goal-directed, reactive, endowed with (bounded) rationality, interactive, mobile, adaptive, with a form of memory or learning;

  - *Interdependence*: agents influence others in response to the influence that they receive, or indirectly through modification of the environment.

Also, and most importantly, the epistemology of agent-based simulation allows for a different relation between implementation and model, e.g. in the case of studies in robotics, where the role of the model with respect to the target system is no longer descriptive or predictive but most often exploratory.[18]

- *Multiscale simulations*: these simulate systems describing elements that live at different scales: for example, a model simulating the stress on matter but focuses on specific regions of the material where important small-scale effects are taking place, while it models those smaller regions with relatively more fine-grained methods.

---

[18] For an extensive analysis of this argument, see e.g. Primiero (2019b). For a more detailed qualification of the distinction between explanatory, predictive, and exploratory aims in experimental computing, see Chapter 16.

Multiscale simulation methods allow for serial and parallel formulations: serial multi-scale modelling simulates every single lower-level area of description and summarizes the results algorithmically at the higher level; the simultaneous simulation of different regions proper of parallel multi-scale modelling is required when the observed behaviour results from the interaction of those different submodels.

- *Monte Carlo simulations*: implementations of computer algorithms for stochastic problems whose solution can be obtained by random number selection mimicking the random behaviour of the system of reference, or for deterministic problems without random behaviours whose solution can be obtained as the average result of a stochastic model over many trials. The general structure of this method consists in obtaining an estimate, a standard deviation and its reduction. An example of this type of method is the numerical solution by Ulam and von Neumann to the problem of determining the reaction of a neutron to the penetration of matter, essentially consisting in following its progress in passing the matter, estimating the distance to the nucleus and deciding the behaviour by random choice when getting close to it. Another example is the random-walk problem, of establishing how the final distance from the starting point of a walk of n steps each of length d in random directions depends upon n.

14.3 Epistemic Role of Computer Simulations

The complex relation between computer simulations and experiments, and the possibility of qualifying the former ones as a kind of the larger class of experimental activities has been explored at length, as well as their epistemological nature in relation to laboratory practices.[19]

There are several possible ways of interpreting the epistemic role of simulations, depending on the type of methodological assumptions, field of study, and scientific aims under which the relation is analysed. Let us start from an overview of these positions, in the standard assumption that the modelling exercise relies on the assumption of the existence of the target system, hence as a form of target-directed modelling.[20]

Simulations present an essential characteristic of experimental computing, and in this they crucially differ from analytical model resolution: simulations combine the very core of the three levels of abstraction of computing we are investigating, namely the formal, the linguistic, and the physical. Simulation is in fact the result of a formal model design, its translation to an algorithmic encoding, and the performing of computational runs on a

---

[19] See respectively Guala (2002); Morrison (2009); Winsberg (2010), and Barberousse et al. (2009); Tal (2011). For a brief overview of the main topics related to the epistemology of computer simulations, see Parker (2013) and Durán (2013); for a more extensive presentation of problems and approaches, see Durán (2018). For an overview of the literature on simulations and experiments, see Humphreys (1990); Hartmann (1996), and Humphreys (2004).

[20] See (Weisberg, 2013, ch.5).

physical device. As such, the epistemic status of computer simulations reflects the whole complexity of experimental computing:

> An obvious characterisation is that [computer simulation] is some type of hybrid activity that involves aspects of modelling and experimentation and as such occupies a middle ground. The result is that data produced via simulation have an epistemic status that is less robust than that of experimental data but greater than that of the results obtained from the mathematical manipulation of or calculation with models.[21]

Simulation in this sense can refer both to a computer program simulating the target system of interest, as well as another physical implementation (e.g. in an artificial or natural material)[22] which is believed to have a good model with fundamental mathematical similarities to the model of the systems of interest. Observations of the former then allow inference to be made about the latter. Obviously only simulations in the form of computer programs reflect the type of critical layered structure that includes the formal model, its algorithmic and linguistic translation, and the physical implementation that is of interest to us in the present context. Hence, our aim is now to overview the epistemic role specific to computer simulations. Three main approaches are identified:

1. the epistemic role of simulation can be defined as to reflect the analysis made for models in Section 14.2. Simulation can be used to understand some current or past behaviour of a given system based on available data, to explain how events (could) have occurred or properties be obtained. In this case we attribute to simulations an *explanatory power*.

2. simulations can be used to discover new information about the target system, to predict data that we do not have, to predict the behaviour of a system in the real world under certain given circumstances, or to predict future states of the system. Predictions are characterized as point, system, or range predictions, in which states or properties of the system of interest are identified respectively at a specific point in time, as general and stable properties of the system or as range variations over currently known values within certain physical or temporal boundaries. In these cases, we attribute to simulations an heuristic role or *predictive power*.

3. simulations allow us to represent information to ourselves or communicate it to others and in doing so they make any system, even one under construction, knowable. In this sense simulations display an *exploratory power*. This latter aspect of experimentation is less considered in the literature and this epistemic character-ization of experiments is best suited to their qualification not just with computa-tional aids but especially for their use in computational domains, like autonomous robotics, empirical software engineering, industrial design. Explorative experi-ments are characterized by a lower degree of constraining from the control part of

---

[21] (Morrison, 2015, p.248).
[22] See e.g. Trenholme (1994).

the activity, often without the guide of a formal model, and driven by the interest in verifying the possibilities of an artefact. In this sense, it is the experiment that guides the construction of the computational model, or at least contributes to it. In this process, a computational hypothesis is not conceptually prior to the design and execution of the computational experiment: accordingly, control is *a posteriori*, i.e. it is performed only at a much later stage and often contextually to the artefact's use. The most important characterization so far of this type of experiment is that it is *directly action-guiding*, i.e. there is a goal of human action to be attained, and the interventions studied aim at being performed in a non-experimental setting to achieve that goal.[23]

This essentially novel understanding of computational experiments as exploratory experiments is most apt when the area of investigation is a computational science, and it is still less developed and investigated than its explanatory and predictive counterparts. A contributing factor in its conceptual definition is the definition of an appropriate notion of simulation specifically for the artificial sciences. This definition relies on the view that simulation contributes to the design of a computational model used to formulate the reality of reference; that verification and validity strongly differ in their definition from the notion of simulation in the standard sense; and that exploration contributes to problem-solving rather than explanation or prediction. This leads to the major task of defining formal relations between simulation, computational, and formal models for all different types of experimental computing, namely explanatory, predictive, and exploratory. This task is reconsidered at length in the next chapters.

Exercises

Exercise 135 *Describe the different types of models.*

Exercise 136 *Describe briefly the essential steps of the modelling practice.*

Exercise 137 *Explain the main approaches to the relation between models and target systems in terms of laws.*

Exercise 138 *Describe the different epistemic roles of models.*

Exercise 139 *What are the possible conditions under which models and target systems relate?*

Exercise 140 *Explain the broad notion of simulation.*

Exercise 141 *Explain the narrow notion of simulation.*

Exercise 142 *Offer an overview of simulation methods.*

Exercise 143 *Describe the different epistemic roles of simulations.*

---

[23] See (Schiaffonati, 2016, p.656).

15 Formal Relations

Summary

In this chapter, we investigate further in formal detail the epistemic properties of the relations between computer simulations, models, and their target system. We review current analyses present in the relevant literature and advance our view on some formal methodological aspects required for a graded approach to such relations.

15.1 Identity and Dependence

A first approach to the foundational discussion on the epistemology of computer simulations argues against its novelty, articulated by rejecting the following points:[1]

- *Metaphysical Claim*: Simulations create some kind of parallel world in which experiments can be conducted under more favourable conditions than in the real world;
- *Epistemological Claim*: Simulations demand a new epistemology;
- *Semantic Claim*: Simulations demand a new analysis of how models and theories relate to concrete phenomena;
- *Methodological Claim*: Simulating is a *sui generis* activity that lies 'in between' theorizing and experimentation.

According to this view, simulations do not offer more favourable conceptual conditions and more valid results than experiments; accordingly, simulations do not need to be explained and guided methodologically in any different way than standard scientific

---

[1] See Frigg and Reiss (2009).

On the Foundations of Computing. Giuseppe Primiero, Oxford University Press (2020). © Giuseppe Primiero.
DOI: 10.1093/oso/9780198835646.001.0001

enterprises; they do not require a different interpretation of the relation between model and theory, and their methodological nature is no more complex than what scientific practice knows from the standard theory–experiment relation. In this sense, computer simulations offer, from the epistemological viewpoint, *no novelty* when compared to standard experimental practices in the non-computational sciences. This view has been explained in the following terms:

> Inasmuch as the simulation has abstracted from the material content of the system being simulated, has employed various simplifications in the model, and uses only the mathematical form, it obviously and trivially differs from the 'real thing', but in this respect, there is no difference between simulations and any other kind of mathematical model, and it is primarily when computer simulations are used in place of empirical experiments that this element of falsity is important.[2]

But the epistemic role of simulations in terms of their explanatory, predictive, and exploratory power seems to rely on the different contexts and uses of simulations and their models in scientific practices, and this seems to oppose the above-reported view that simulations and experiments do not differ from a methodological and epistemological viewpoint. In the following, we consider this debate from the point of view of the formal relations between target system, model, and simulation.

Assuming the ability of experiments to reflect a faithful construction of the model of the target system, the relation between the computational model underlying the simulation and the corresponding mathematical or theoretical template abtracted from reality is assumed to be one of perfect mimesis.[3] This position is usually formulated in terms of the following two theses:

Thesis 14 (Identity Thesis) *Computer simulations are instances of experiments.*

Thesis 15 (Epistemological Dependence Thesis) *If computer simulations are instances of experiments, then they can provide a guarantee to believe in their results.*

Assuming some form of the identity thesis, the meaning of the Dependence Thesis is that simulations guarantee the same type of epistemological support for the beliefs they induce as experiments do. In other words, a simulation is then experimentally valid for the target system. The crucial element in this theoretical position is represented by the assumption that there is a strict identity relation between the target system, its formal interpretation, a computational translation, and eventually its implementation in the simulation: if the conditions of identity of these elements are met correctly, then a simulation can be used to test or detect a real world phenomenon. We can now express this relation as a property between models and their implementations in some computational artefact like a simulation; see Figure 15.1.[4]

---

[2] (Humphreys, 1990, pp. 501–2).

[3] (Humphreys, 1990, pp. 499–500).

[4] In this figure and in all the following ones in the present chapter and in Chapter 16, we include the term 'Artefact' in the bottom-left corner to refer to a computational implementation of a model to run experiments:

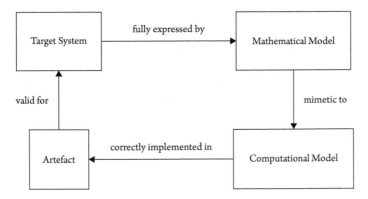

Figure 15.1 Identity relations between system, formal model, computational translation, and physical implementation

Unfortunately, it is not unproblematic to decide when these elements meet the right conditions to satisfy identity relations. As the relata stand at different levels of abstraction, assuming an identity relation requires:

1. determining that an exact mapping holds between target system and formal model;
2. that a similar relation holds in the translation to a computational model;
3. and finally that the implementation is fully correct with respect to the computational model.

Moreover, from the practical point of view, simulations are notoriously treated as idealizations or abstractions on reality: if one assumes that an abstraction cannot be—by definition—an exact mapping of real properties and relations, it seems that a weakening of the identity relation is necessary. In turn, simulations cannot simply be assumed to require the same epistemic treatment—or provide the same level of justification—as standard experimentation:

> But if the underlying mathematical model can be realistically construed (i.e. it is not a mere heuristic device) and is well-confirmed, then the simulation will be as 'realistic' as any theoretical representation is. Of course, approximations and idealizations are often used in the simulation that are additional to those used in the underlying model, but this is a difference in degree rather than in kind.[5]

If one rejects the thesis that implementations of models in computational artefacts like simulations are (in some sense) equivalent to real experiments, then the real-world

---

this is the narrow sense of simulation considered in Frigg and Reiss (2009) and reported in Chapter 14.2. The picture as a whole can be understood as a representation of the broad sense of simulation.

[5] (Humphreys, 1990, pp. 501–2).

system object of analysis is accounted for by such artefacts in a different way. One useful operation is then to understand which weaker types of relation between the model and its implementation can be defined. The standard epistemic analyses have considered the following possible relations:

- *isomorphism*: every property, element, or relation of the real-world situation is mapped to a formal property in the artefact;
- *analogy*: every property, element, or relation in the artefact is a simplified version of a property of the real world;
- *similarity*: the artefact includes properties, elements, or relations with similar counterparts in the real-world system.

Several technical, social, and conceptual aspects invest the problem of determining which of these relations occur between implementation (e.g. a simulation) and the object of the model (e.g. the target system). Our focus, in the light of an experimental foundation of computing, is to explore these relations from a formal point of view in order to establish whether and how implementations (and computer simulations in particular) can be characterized as experiment-like activities.

15.2 Isomorphism

A moderate epistemological reading of the relation between standard scientific knowledge and practice based on computational artefacts like simulations relies on the idea that an isomorphism can be established between the formal model and the system. When this is the case, and assuming the usual constraints on precision of the computational translation and correctness of the implementation, there is a map between the properties and relations of the artefact and those of the target system; see Figure 15.2.

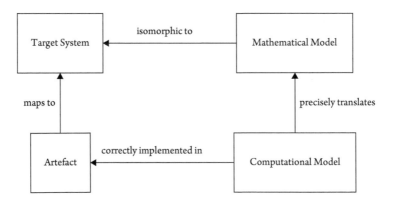

Figure 15.2 Isomorphic relations between system, formal model, computational translation, and physical implementation

An isomorphic relation between simulation and the simulated process is a property-preserving relation between distinct structures of reference:

Definition 139 (Isomorphism) *An isomorphism (or bijective morphism) between A and B is a map that preserves sets and relations among elements.*

The characteristics of an isomorphism—when compared to identity—is that the former can preserve properties through translation of different structures: for example, a logarithmic function from the multiplicative group of positive real numbers is isomorphic to the additive group of real numbers. When the notion of isomorphism is applied to the system-model-implementation relation, the physical process object of the scientific investigation maintains a conceptual priority over the implementation: this priority is required because the isomorphism works by translating observable behaviours of the target system to computational processes in the computational artefact, e.g. in language-specific properties of the simulation.

An example of such a weaker identity relation between model and target system is based on a characterization of simulations as isomorphic but not identical to experiments with respect to the following aspects:[6]

- *Visualization*: the process of setting up an experiment in a standard scientific setting and *observing* the behaviour of the system under given initial conditions is comparable to the process of using visualization techniques for simulation systems dealing with massive amount of data (number of agents, environment conditions, and so forth);

- *Approximation*: distortions are true of any scale model and, more generally, of any physical system not strictly identical to the target system; in this sense, a simulation approximates the reality of the simulated system in a manner comparable to the approximation of experiments in the standard scientific practice;

- *Discretization*: in a real-world experiment, both the experimental and target processes may well both be continuous processes, but the experimenter will use them (in either manipulation or observation) only with some finite degree of error;

- *Calibration*: calibration in simulation serves the same purpose as in a physical experiment, to find the settings supporting previously observed measurements of a target system under given initial conditions.

These aspects shared by simulations and experiments are related to the process of *verification*, i.e. the act of determining whether the simulation correctly implements the theory being investigated, requiring processes like design verification, debugging, and consistency checks.[7] According to this view, computer simulations can be considered a proper method of experimentation, but in doing so they present new problems related to

---

[6] For such a view, see e.g. (Korb and Mascaro, 2009, Section 6). These aspects are also referred to in (Primiero, 2019b, sec.2)

[7] See (Korb and Mascaro, 2009, sec.4.2).

the techniques in use, limited by computability theory rather than by physical limits. In this sense, their relation is a mapping, rather than an exact match, to the target system.

This analysis obviously ends up insufficient for those contexts in which computational artefacts, e.g. simulations, do not play a purely explanatory role, and are instead characterized either as predictive or as exploratory.

15.3 Analogy and Similarity

A further weakening of the identity thesis relies on an understanding of the relation between models and implementations as requiring true novel properties when compared to standard experimentation. The epistemic difference between the formal model on which the implementation relies and the target system can be so strong that it might qualify as simply analogical. This has been expressed specifically for computer simulations:

> a simulation is any system that is believed, or hoped, to have dynamical behavior that is similar enough to some other system such that the former can be studied to learn about the latter.[8]

This position usually takes into account the following main problems related to the epistemological status of simulations:[9]

- *Epistemic Opacity*: a process is epistemically opaque relative to a cognitive agent X at time t just in case X does not know at t all of the epistemically relevant elements of the process; this is to say that within simulation a specific set of variables is chosen and the methodological validity of the process is confined to such a limited set of elements, other aspects and their influence remaining inaccessible to the investigation;

- *Semantics*: the way in which simulations are applied to systems is different from the way in which traditional models are applied: while the latter are required to denote the model of reality under analysis, in simulation the relation is less rigid and proceeds rather by approximation;

- *Temporal Dynamics*: while in a traditional scientific setting one requires a temporal representation of the dynamical development of the system under observation, in the case of simulations there is additionally a temporal process involved in actually computing the consequences of the underlying model, thus inducing a different, over-imposed temporal dynamics;

- *Practice*: finally, the computational setting in which simulations occur, illustrates a separation between what can be computed in theory and what can be computed with the available resources; this aspect must be considered also in the opposite

---

[8] (Winsberg, 2015, sec.1.3).
[9] See e.g. (Humphreys, 2009, p.625).

direction, with computational means allowing more than what the system of reference can.

With respect to the problem of epistemic opacity, the level of knowledge that the cognitive agent X can exhibit with respect to the epistemologically relevant elements of the process can vary, depending on the level of access and competence that X has with the implementation, and in particular depending on whether X is the designer, the programmer, or simply the user. This analysis can be better formulated by qualifying which level of access is granted to which agent. From the semantic point of view, the gap between system of reference, model, and implementation suggests that a layer of complexity is added by simulations being different technical artefacts than their models: in this case, our analysis should consider whether the implementation is posterior to the model and whether the relation is one of isomorphism, analogy, or just similarity. For the temporal characterization, a simulation compared with an underlying (theoretical) model, for which it acts as inferential engine, will import a different notion of time and the relation between the temporal representation of the process of interest at the two levels needs to be addressed. The last aspect, concerning the distinction between applicability in practice and in principle must be considered in view of the design and the implementation.

These problems, it can be argued, arise because simulations hold an epistemic interaction with models of the events rather than with events themselves, as is the case with experiments.[10] Other authors supporting the analogical relation between simulation and target system require the experimental practice to involve modelling,[11] and they argue that simulations manipulate objects that only formally resemble the target, while the experiments manipulate objects with a material similarity with the target.[12] In simulation, it seems, the object of interest is always distinct from the object investigated (the computing system performing the simulation), and this represents a crucial difference from standard experiments.[13]

The view that simulations, along with any computational artefact used for experimental purposes, consist of manipulation of abstract objects has been criticized because it is difficult to establish the meaning of such an operation, or because standard controlled experiments are nothing different: external validation, i.e. establishing that the experimental results actually apply to the system of interest, is a true epistemic hurdle, and it seems that this holds for any computational artefact in relation to the model it implements.[14] To summarize: while the formal model holds a similarity relation with the target system, the artefact maps to the model and this grounds the analogy relation to the target system; see Figure 15.3.

The problems associated with the analogical view on the relation between computational artefact and target system can be formulated as follows:

---

[10] See (Gilbert and Troitzsch, 1999, p.13).
[11] See Morrison (2009).
[12] See Guala (2002).
[13] For a full summary of this debate, see Winsberg (2015).
[14] See Guala (2002, 2008); Morgan (2003); Parker (2009); Winsberg (2009).

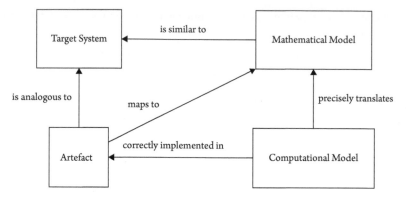

Figure 15.3 Analogy relations between system, formal model, computational translation, and physical implementation

1. it is impossible to determine precise properties of validity and verification;
2. it is impossible to qualify classes of implementations valid for a given target system;
3. it is impossible to establish composed systems with corresponding compositions of models.

The existence of a formal similarity between the artefact and the system investigated is not sufficiently clear to establish which relation holds, in turn, between the formal model and the target system.[15] There are many ways in which formal identity or similarity can be shown. Moreover, a formal relation of this kind holds only among representations of those entities rather than among entities themselves. This should suggest that a formal relation between such descriptions could be defined: the aim of Section 15.4 is to provide such an account of formal relations between the computational description of the target system and its formal description.

15.4 Variants of Simulationism

The analysis of the previous sections has illustrated the problems associated with existing views on the relations between formal and computational models, implementations, and target systems. In the following, we propose to abstract entirely from the problem of determining the relation between model and target system, and to analyse possible formal relations between the mathematical and the computational model of the intended target system. These are well-defined structures for which appropriate formal relations can be formulated, in turn allowing one to establish epistemic properties associated with verification and validation for the implementation.

---

[15] See Winsberg (2015).

To formulate our approach, let us start by considering two main positions in the debate about the computational description of physical systems:[16]

1. *Weak Simulationism Principle*: if a physical system is exactly simulated by a computation, then that type of computation is all there is to the nature of the physical system.

2. *Strong Simulationism Principle*: every physical system can be simulated by a physical computation.[17]

The simulationists' principles offer an approach to formulate formal variants which can be used to establish precise relations between entities or descriptions thereof. In particular, we can formalize the relation between an abstract, mathematically defined description of a physical system and its computational description (including its implemented algorithmic translation) in terms of the simulationist principles. In order to do so, we assume that the remaining relations between the entities involved are strong ones, and in particular that the mathematical or formal model is at least isomorphic with respect the target system, and that the implemented artefact of interest (e.g. the simulation) is correct with respect to the algorithmic description of the system. Our focus is then on the novel relation between the mathematical model of the target system and its algorithmic translation. We claim that establishing formal relations between mathematical and computational models allows a direct relationship to be restored between the implementation of the latter and the target system.

Isomorphism relations are deficient in formulating the relations between models and target system because of their limited ability to account for partiality and idealization. These problems have been accounted for in terms of partial isomorphic-based relations.[18] Let us recall that in Definition 78 from Chapter 6 the formal relation of simulation between two systems implementing the same computational runs was introduced. Definition 79 extends this notion to a relation holding in both directions between the two systems.[19] These formal relations can be subsumed under the epistemic principles of weak and strong simulationism.

Principle 18 (Weak Simulationism) *A weak simulationist relation holds between two computational structures if one can fully simulate the other, but the simulation can be incomplete in the other direction.*

---

[16] See Chapter 10 for the philosophical debate on physical computation. For these views, see e.g. Piccinini and Anderson (2018).

[17] The strong version supports pancomputationalism, namely the thesis that the universe itself is a computational simulation run by a physical computer that exists in a separate physical universe. As we are not interested in (either weak or strong) pancomputationalist views, we reformulate these principles here in a way that is independent of their impact on the ontology of the universe.

[18] See e.g. Bueno et al. (2002) and da Costa and French (2003). Our account can be seen as a further improvement on such approaches with clearly defined properties, aiming at answering some of the still existing limitations to structuralist approaches.

[19] The former can be used to interpret the notion of inexact copy between computational artefacts, the latter relation interprets instead the notion of exact copy. See Angius and Primiero (2018) for these translations in terms of equivalent notions for state transitions systems.

Mathematical Model	Computational Model
Finite Description of Temporal States	$S(\mathcal{A})$
Names of Functional Transformations	$\tau_A : S(A) \to S(A)$
States Predicates	State Labels in $\mathcal{A}$

Figure 15.4 Correspondence between mathematical and computational models

Principle 19 (Strong Simulationism) *A strong simulationist relation holds between two computational structures if either system exactly mirrors the other.*

In order to apply these relations, the computational model can be understood as a Sequential Time Algorithm as by Definition 73 matching the formal model of the target system: computational states express (temporally defined) states of affairs of the physical system; the finiteness condition on the set of states can be interpreted as saying that at every given moment in which the interpretation of the physical system by a model is considered, the set of states interpreted by the model is finite; maps between states in the computational model express functional operations transforming states as input into states as output, matching physical transformations of the formal modelling of the target system; state labels are predicates defining properties at states. See Figure 15.4.

The complete description of the mathematical model (MM) of a target system through a description of its static and dynamic properties, the states and transitions allowed by the model, the initial states, and the dependencies between states, is also referred to as the *total state* of the model.[20] The computational translation of MM (here denoted as CM) is then the Sequential Time Algorithm (or any other equivalent formulation) translation of that model, assuming that its implementation is correct with respect to such computational description. Then it is immediate to adapt the notion of simulation to a weak simulationist relation between a physical system and its computational description:

Definition 140 (Weak Simulationism) *Given a mathematical model MM isomorphic to a given target system expressed by a finite set of states S and transitions $t : S \to S$ and given its computational model translation CM expressed by an algorithm $\mathcal{A}$ with a finite set of states $S(\mathcal{A})$ and mappings $\tau_A : S(A) \to S(A)$, CM weakly simulates MM if and only if*

- *for each initial state s_0 of MM there is an initial state $s'_0 \in S(\mathcal{A})$ in CM such that s_0 and s'_0 satisfy the same property; and*
- *there is a successor state s'_1 of s'_0 according to τ_A in CM for each successor state s_1 of s_0 according to t in MM such that they have the same property.*

Under this reading, if a computational description weakly simulates the formal description of a target system, then the former must include an appropriate translation for every state, and mirror correctly every state transition contained in the latter. Note that the computational model might include states and transitions that are not properly part of

---

[20] See (Weisberg, 2013, p.79).

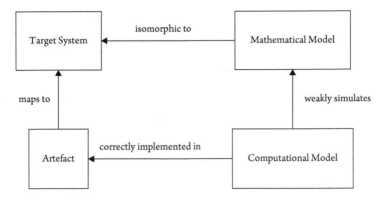

Figure 15.5 The computational model weakly simulates a formal model of the target system

the formal description of the physical system, while it must completely describe it. The relation of weak simulationism $\rightsquigarrow^W$ is then expressed as follows:

$$CM \rightsquigarrow^W MM \leftrightarrow \forall s_0 \in MM \exists s_0' \in S(\mathcal{A}).P(s_0) \equiv P(s_0') \wedge$$
$$\forall s_1 \in MM.(s_0 \rightarrow^t s_1) \exists s_1' \in S(\mathcal{A}).(\tau : s_0' \rightarrow s_1').P(s_1) \equiv P(s_1')$$

If a weak simulation holds between the computational and the formal model of the target system, then this relation assures that a correct implementation of the computational description will include only properties and behaviours expressed by the target system; see Figure 15.5.

Note that the process of defining a formal model, translating it to a computational or algorithmic description, implementing the latter, and then analysing its relation to a target system is often a process based on a feedback. The target system most likely returns properties that direct the implementation design and accordingly correct the models. This is especially true when the target system is artificial, thereby allowing tinkering and testing. In general, the model-oriented process of explanation will feedback to the design by identifying simplifications of the theoretical model which are discovered and possibly corrected in the computational translation. In these cases, it is possible to design an artefact which implements the computational model and expresses the properties of the mathematical or formal model; see Figure 15.6. We will see in Chapter 16 how the direction of a simulation relation between mathematical and computational models is relevant.

The stronger version of the same relation can be formulated as a symmetric counterpart to weak simulationism:

Definition 141 (Strong Simulationism) *Given a mathematical model MM of a target system and its computational translation CM defined as above, a strong simulation holds between them if and only if CM weakly simulates MM, and vice versa.*

If a strong simulation relation exists between a formal description of a given target system and its computational translation, then for every state in either description there

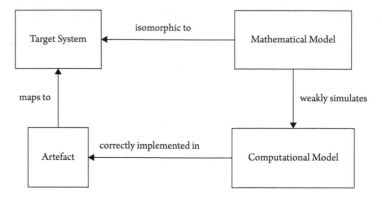

Figure 15.6 The formal model weakly simulates a computational model of the target system

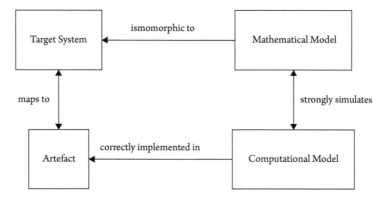

Figure 15.7 A bisimulation exists between formal and computational model of a target system

exists a corresponding state in the other, and for every transition formulated *also* by the computational system, the formal model of the physical system will have an appropriate instance; see Figure 15.7.

$$CM \leftrightsquigarrow MM \leftrightarrow CM \rightsquigarrow^{W} MM \wedge MM \rightsquigarrow^{W} CM$$

The relation of approximate simulation can be formally expressed by introducing a partition on the set-theoretic structure formally defining a system, thereby identifying all behaviours valid from its initial states given the available transitions, and defining a formal simulation on one term of the obtained equivalence relation. By introducing the notion of path π as the sequence of states expressing a behaviour of the system, we can then formulate an approximate simulation relation between the formal and the computational description of a system:[21]

---

[21] See Korb and Mascaro (2009) for the use of this relation to account for both explanation and prediction and Angius and Primiero (2018) for its application to the notion of approximate copy.

Definition 142 (Approximate Simulationism) *Given a a mathematical model MM of a target system and its computational translation CM defined as above, CM approximately simulates MM if and only if*

1. *for at least one path π from an initial state s_0 to a final state s_n in MM according to appropriate transitions $t, t', \ldots$, there is a path $\pi' \in CM$ with an initial state $s'_0 \in S(\mathcal{A})$ to a final state $s_n \in S(\mathcal{A})$ in CM according to appropriate maps $\tau_A, \tau'_A, \ldots$ such that s_0 and s'_0 satisfy the same property and there is a successor state $s'_1 \in S(\mathcal{A})$ of $s'_0 \in S(\mathcal{A})$ according to τ_A in CM for each successor state s_1 of s_0 in π of MM according to t such that they have the same property; and*

2. *there is at least one path $\rho \in MM$ for which there is no path $\rho' \in CM$ such that this relation holds.*

If an approximate simulation relation exists between a computational and a formal description of a target system, then the former simulates at least one behaviour of latter, but there will be behaviours of either not matching behaviours in the counterpart description:

$$CM \rightsquigarrow^A MM \leftrightarrow \exists \pi = \{s_0 \rightarrow^{t_n} s_n\} \in MM \wedge$$
$$\exists \pi' = \{\tau : s'_0 \rightarrow s'_n\} \in CM \wedge$$
$$\pi' \rightsquigarrow^W \pi \wedge$$
$$\exists \rho \in MM. \forall \rho' \in CM \rho' \not\rightsquigarrow^W \rho$$

In the case of an approximate simulation between the computational model and the formal model of the target system, a correct implementation of the former is assured to map to a partial description of the target system; see Figure 15.8.

The simulationist approach offers therefore graded versions of the relation between models. Similarly to what was investigated in Parts I and II of this volume, we are

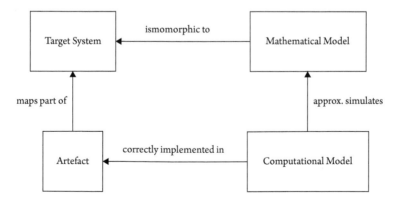

Figure 15.8 Approximate simulationism relation between system, formal model, computational translation, and physical implementation

interested in formulating correctness and validity principles appropriate for the experimental foundation: this will be obtained by considering these notions with respect to the strong, weak, and approximate simulation relation formulated above between the levels of abstraction of interest. This is the aim of Chapter 16.

Exercises

Exercise 144 *Explain how the relation between simulations and experiments is presented in the literature.*

Exercise 145 *Illustrate the thesis that considers simulations identical to experiments from an epistemological point of view.*

Exercise 146 *Explain in which way the identity relation between experiments and simulations can be weakened in terms of an isomorphism.*

Exercise 147 *Illustrate the thesis that considers simulations analogical to experiments from an epistemological point of view.*

Exercise 148 *Explain the Weak and Strong Simulationist Principles and how they relate to the debate about the epistemic value of simulations.*

Exercise 149 *Explain how the Weak Simulationist Principle can be translated as a formal relation between the mathematical model of a target system and its computational translation and which type of relation it induces between simulation and target system.*

Exercise 150 *Explain how the Strong Simulationist Principle can be translated as a formal relation between the mathematical model of a target system and its computational translation and which type of relation it induces between simulation and target system.*

Exercise 151 *Explain how an approximate simulationist relation expresses a formal relation between the mathematical model of a target system and its computational translation and which type of relation it induces between simulation and target system.*

16 Computing as an Experimental Discipline

Summary

In the previous chapters we have extensively considered experimental computing in terms of the possible relations between target system, formal, and computational models and their implementation. In this chapter we consider directly the formal and epistemic properties characterizing computing as an experimental discipline and reconsider validity and correctness under our interpretation.

16.1 A Balanced Approach

Within the approaches to computing as an experimental discipline, the debate has touched both extremes: on the one hand, there are supporters of computing as a discipline radically different from other sciences like physics, for example because its experiments aim at uncovering practical issues with theoretical work rather than refuting theories;[1] on the other hand, there have been approaches that consider computing explicitly as an experimental science, characterized in all its aspects similarly to more standard hard sciences based on the hypothetical-deductive method and supported by the falsification process proper of experimental sciences.[2] There are also positions advocating a balanced approach, supporting the dual nature proper of a discipline with a theoretical foundation, but whose applied nature requires the support of experiments, and with a very specific interpretation of the relation between world, models, and implementations, due to the layered structure of its ontology and epistemology. In this context, experimentation is considered beneficial to a reliable knowledge base construction, guaranteeing uncertainty

---

[1] See e.g. Hartmanis (1994); Hartmanis and Lin (1992).
[2] See e.g. Plaice (1995).

On the Foundations of Computing. Giuseppe Primiero, Oxford University Press (2020). © Giuseppe Primiero.
DOI: 10.1093/oso/9780198835646.001.0001

reduction concerning theories, methods, and tools. Moreover, experimentation is deemed essential to help knowledge construction in areas where engineering is slow.[3] It is in the spirit of a discipline aiming at the progress of knowledge to maintain a balance between its speculative and its experimental nature.[4]

A further set of arguments in favour of the balanced approach can be proposed, based on the ontological nature of computing. Computing as a discipline construed around several levels of abstraction (from hardware up to intention) has been defined as the study of both ontologies and epistemology of information structures. As such, computing is in itself projected both towards a more abstract and formal apparatus (the epistemology) as well as physical and empirical infrastructures (the ontologies). In this sense, it is obvious that a methodological study of the discipline requires an equilibrium between theory construction and theory verification, where the former guides the latter and this in turn re-addresses the former. We have seen such dynamics at play already in the definition of the problem of correctness, with its formal analysis based on the proofs-as-programs isomorphism in Chapter 7. We have also considered the principles guiding its physical counterpart (functionality, efficiency, usability) and to some extent the role of testing in Chapter 10. We have moreover anticipated how the distinction between explanatory, predictive, and exploratory power of computational experiments (and simulations in particular) reflects this dynamic relation between models and implementations.

It is now essential to illustrate how the experimental nature of computing is influenced and influences the design of theories and which notions of validity and correctness are admissible by the experimental foundation of computing. So far in Part III of the volume, and in Chapter 15 in particular, we have presented our view on the relation between computational models and computational artefacts, implementing them as a way of providing an experimental basis. In this chapter, we provide an appropriate epistemological analysis to designing and exploring computing systems for experimental purposes in view of their verification and validation properties.

16.2 Evaluation

The evaluation of computational artefacts intended as experimental tools through implementation of corresponding models can be done by borrowing the analysis known from the standard epistemology of computer simulations, which relies on a two-steps process:[5] verification considers the relationship between implementation and specification; validation focuses on the relationship between the model and the target system. Our interest is thus now on how this combined process can be defined on the basis of the formal relations formulated in Chapter 15. The introduction of possible isomorphisms and weakening thereof between the formal and the computational models of the target system gives a

---

[3] See e.g. Tichy (1998).
[4] See Génova (2010).
[5] See e.g. (North and Macal, 2007, pp.30–1) and (Durán, 2018, pp.109–10).

chance to provide fresh formulations for these essential epistemic properties. This, in turn, can shed light on the validity and correctness notions for experimental computing in general.

Verification

Let us start with verification. Methods to establish it are usually considered in terms of correctness of code implementation, correctness of numerical algorithms, and practical correctness of simulations in terms of execution.[6] The assumption of a weak simulationist relation between the computational and the formal or mathematical model induces a different degree of epistemic certainty concerning the relation between the implementation and the specification. We understand the latter to be the relation holding between the implemented artefact *IM* and the mathematical model *MM*. If a weak simulationist relation holds between computational *CM* and mathematical model *MM*, a correct implementation of the former can only weakly verify the latter (see Figure 16.1):

Definition 143 (Weak Verification) *A weak verification is satisfied between an implementation IM and its specification in terms of a mathematical model MM if and only if every computational trace π of IM matches a behaviour ρ in MM:*

$$IM \leadsto^{wver} MM \leftrightarrow \forall \pi \in IM \exists \rho \in MM.\pi \equiv \rho$$

Note that if a weak simulationist relation holds of the computational model by the mathematical model according to Figure 15.6, and preserving correct implementation, the relation of weak verification remains unchanged.

If a strong simulationist relation holds between the computational *CM* and the mathematical model *MM*, then it is immediate that every property and transition in the model must be accounted for in the implemented artefact *IM*. This corresponds to an

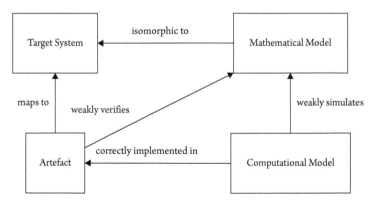

Figure 16.1 Weak verification

6 See (Durán, 2018, pp.110–11).

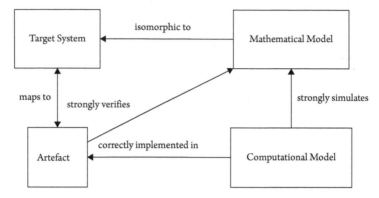

Figure 16.2 Strong verification

appropriate relation of verification in terms of the total state of the mathematical model (see also Figure 16.2):

Definition 144 (Strong Verification) *A strong verification is satisfied between an implementation IM and its specification in terms of a mathematical model MM if and only if every computational path ρ in the latter occurs as a behaviour π in the former:*

$$IM \leadsto^{sver} MM \leftrightarrow \forall \rho \in MM \exists \pi \in IM.\rho \equiv \pi$$

If an approximate simulationist relation holds between computational and formal model, there is at least one property and transition in the model which is accounted for in the implemented artefact, but the mapping is not exhaustive (see also Figure 16.3):

Definition 145 (Approximate Verification) *An approximate verification is satisfied between an implementation IM and its specification in terms of a mathematical model MM if and only if at least one partition ρ of the total state of the latter occurs as a computational trace π in the former, while at least one is not:*

$$IM \leadsto^{aver} MM \leftrightarrow \exists \rho \in MM \exists \pi \in IM.\rho \equiv \pi \wedge \exists \rho' \in MM.\neg \exists \pi' \in IM.\rho' \equiv \pi'$$

Validation

So far, we have assumed that the mathematical model is isomorphic to the target system. Now we want to qualify this relation on the basis of the verification process illustrated above. This brings us to validation. Validation methods are usually explained in terms of comparison with experimental measures, extrapolation of conditions of intended use of the model, estimation of accuracy of the model with respect to specified requirements.[7] From a formal point of view, the existence of a weak simulationist relation between the

---

[7] See e.g. (Durán, 2018, pp.111–12).

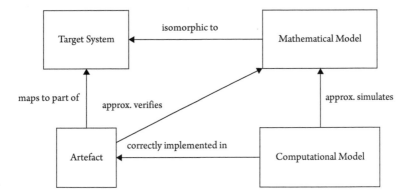

Figure 16.3 Approximate verification

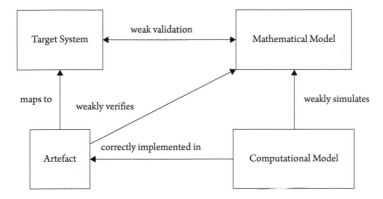

Figure 16.4 Weak validation

computational and the formal model of the target system, which induces weak verification of the latter by the implementation, expresses weak validation (see Figure 16.4):

Definition 146 (Weak Validation) *Weak validation is satisfied between the mathematical model MM and the target system TS if and only if every behaviour ρ that composes the total state of the former is manifested by a behaviour σ of the latter:*

$$MM \rightsquigarrow^{wval} TS \leftrightarrow \forall \rho \in MM \exists \sigma \in TS.\sigma \equiv \rho$$

The existence of a strong simulationist relation guarantees that every dynamical behaviour illustrated by the target system is matched by its formal model, i.e. there is nothing more and nothing less that the real-world system (intended or real) does when compared to the model (see Figure 16.5):

Definition 147 (Strong Validation) *A strong validation is satisfied between the mathematical model MM and the target system TS if an only if every behaviour σ that composes the total state of the latter is expressed by some behaviour ρ of the former:*

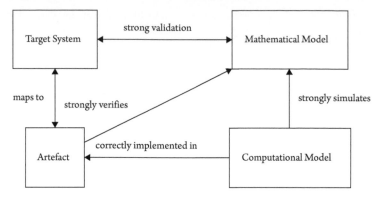

Figure 16.5 Strong validation

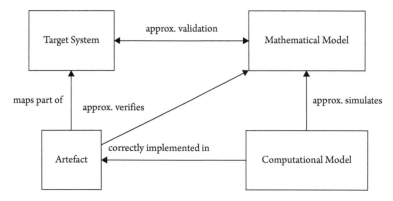

Figure 16.6 Approximate validation

$$MM \leadsto^{sval} TS \leftrightarrow \forall \sigma \in TS \exists \rho \in MM.\sigma \equiv \rho$$

While approximation in the case of verification simply means that only a partial correctness property can be ascribed to the implemented model with respect to its intended design, more interesting is the definition of an even weaker validation relation (see Figure 16.6):

Definition 148 (Approximate Validation) *An approximate validation is satisfied between the mathematical model MM and the target system TS if and only if there is a behaviour ρ of the former which is equivalent to a significant part σ of the target system, while not exhausting it:*

$$MM \leadsto^{aval} TS \leftrightarrow \exists \rho \in MM \exists \sigma \in TS.\rho \equiv \sigma \wedge \exists \rho' \in MM.\neg \exists \sigma' \in TS.\rho' \equiv \sigma'$$

An approximate simulationist relation between an implemented computational model and some part of a target system is a standard situation in the construction of complex models, where one starts from simple models and looks for an integration of those approximations to render or create entire new systems.

Properties

The evaluation of the model-world relation determined by our formal analysis can be assessed in terms of criteria present in the literature:[8]

- maximality: a model is maximally similar to itself and to any target that shares all of its properties;
- scalarity: models can tell a greater or lesser number of true things, depending on their degree of idealization;
- richness: the model-world relation ought to allow for different kinds of features to make comparisons;
- quality: it ought also to allow for qualitative comparisons;
- idealization: it should distinguish more from less successful instances of ideal models;
- contextuality: the evaluation of the model-world relation depends on the context of application;
- adjudication: it should allow for extra-empirical disagreement, i.e. related to context, use, weighting of features;
- contentual tractability: it should allow to be reflected and established grounds for actual judgements by scientists.

A strong simulationist relation between computational and formal model, justifying in turn strong validation of the model with respect to the target system guarantees maximality with respect to any non-identical model which does not instantiate a similarly strong relation:

$$\forall MM, MM', MM'' \forall TS(MM \rightsquigarrow^{aval} TS \leq MM' \rightsquigarrow^{wval} TS \leq MM'' \rightsquigarrow^{sval} TS = 1)$$

Our approach escapes some of the main critiques to structuralist interpretations of the required properties of model-world relations. First of all, the formal approach to approximate, weak, and strong validation subsumes—by definition—a quantitative interpretation. Moreover, it allows (thanks to the introduction of an approximate relation) a scalar representation of the world-model relation. In turn, the abstraction process from implementation to computational to formal model implies that different kinds of idealization are possible by implementing different properties, and thus resulting in different partitions of the target system: the more fine-grained the partition, the more

---

[8] See (Weisberg, 2013, ch.8.1). Note that we have qualified the tractability criterion as a contentual one to distinguish it from computational tractability.

closely relations can be defined to target subsystems. By combinations of different partitions, approximate validations allow for idealized models whose properties result from combinations of computational traces mapping to distinct behaviours, with or without a target system. Finally, these definitions subsume a computational tractability of the modelling process with respect to the target system.

It is harder to account for qualitative features, context, adjudication, and content tractability criteria, as these seem to refer to extra-formal properties and as such they cannot depend on purely formal definitions. In particular, it does not seem that these features are a matter of model definition but rather a matter of interpretation. As such, it seems to be true that any structure with a given level of isomorphism to a model holds an identical relation with it, even if none is interpreted from a content perspective, i.e. independently whether we refer to a city, a population, or a biological system. Another critique concerns the emergence of qualitative features in the target system which cannot emerge explicitly in the model. However, a first characteristic differentiating our approach compared to those based on simple isomorphisms is that our definition based on simulation and bisimulation preserves a behavioural identity rather than a purely structural one; in this sense, we do not only consider structures but specifically how these structures are dynamically formed. In particular, the qualitative features of the target system can be matched by the processes present in the formal system: this guarantees more than a structural isomorphism between model and target system. Contextual properties should be accounted for in terms of the formulation of the state space defining the system underlying the formal model; and the weighting of parameters which can facilitate epistemic adjudication can be obtained as well by different subsets of initial states and extensions in terms of probabilistic weighting of transitions. Hence, also practical tractability is not outside the scope of our approach.

16.3 Maximal Criteria: Robustness and Reliability

Verification and validation are considered the basis for asserting reliability of computational artefacts like simulations.[9] In turn, reliability is a guarantee for the trustworthiness of predictions, and so to justify the use of such artefacts in scientific explanations. The standard epistemological practice requires robust theories to estimate the underlying idealizations and it assumes a certain degree of precision. A robustness analysis aims at determining whether given results are due either to the essential aspects of the theory, or the idealization and assumptions made. Robustness in this context indicates the ability of a theory to reach the same essential results also in view of different simplifying assumptions.[10]

---

[9] Several other epistemic properties are referred to in the literature as constituting grounds for reliability, like social, technological, and mathematical opacity. For these other aspects, see e.g. Kaminski et al. (2018). We are here interested only in the contribution of formal properties to the definitions of robustness and reliability.

[10] For an analysis of robustness, see e.g. (Weisberg, 2013, ch.9).

Different approaches to this problem have been determined by philosophical positions. On the one hand, implementations of models in computational artefacts like simulations are built where there is a need to obtain results from sparse data in order to confirm a theory, and in this role, they are expected to induce trust on the basis of a success criterion: in a realist approach, such criterion can be assumed to be truth for both models and our beliefs in them. But once the principle of identity is upheld in favour of a different relation (e.g. we move from identity to isomorphism to analogy), then computational artefacts and their models seem to require validation independently from a criterion of truth for the target system. Obviously, the analysis in terms of graded simulationist relations will impact this issue as well. Validation for model-building techniques characterizes our approach in terms weaker than truth. The success criteria induced by strong validation and strong verification refer to robustness and reliability as elements of a practical and non-fundamentalist approach to the role that computational artefacts and model-building have in the context of computationally structured sciences.[11]

First, we need to define model robustness as a weaker replacement for truth:

Definition 149 (Robust Formal Model) *A formal model is robust if and only if*

- *it correctly represents all required sets of variables of the chosen level of abstraction of the target system;*
- *it correctly represents all inferential properties of the chosen level of abstraction of the target system;*
- *it ensures validity of properties of any robust extension of the model (i.e. any extension by other properties that are considered to satisfy the same level of correctness).*[12]

The first and second properties (full representation of the correct parameters and inferential structure) are meant to guarantee the standard robustness property, i.e. the ability to predict the results common to any model strongly validating the target system. Note that this accounts for parameter and structural robustness:[13]

1. the selection of a strongly validating structure between formal system and target system tells us directly that this is the minimal relation across the different possible models to give rise to the robustness property;

2. the selection of the correct LoA ensures that parameter changing must be valid within the same core set of parameters for the given LoA, it must extend to the same new variables admissible by the given LoA, and it must terminate at the same extension of non-valid parameters for the LoA under observation;

3. structural robustness is tested by changes to the underlying state transition systems representing the formal system strongly validating the target system and it is

---

[11] See e.g. (Winsberg, 2010, ch.7). For a pragmatic theory of truth suitable for computer simulations under the realist perspective, see instead Hubig and Kaminski (2017).

[12] This latter property of robust models can be assimilated to the robustness property for phenomena as theory transcendence or continuity through theory change in (da Costa and French, 2003, p.72).

[13] See (Weisberg, 2013, ch.9.2).

meant to persist as long as the transitions are mapped in a way that still preserves such relation.

The third property (robust model extension) must be understood as to include model integration, i.e. the ability of a robust model to integrate its model description with those of other models preserving data observation and calculation correctness. Robust model extension and integration can be understood as operations of ordering of distinct levels of abstraction, each constituting (at least) a weak validation of the complex resulting model.[14] Note that a robust model representation of a formal system allows by definition for predictive measurements on the variables of interest to both model and system. The last type of robustness[15] concerns the preservation of attributes and structures across different representations of the model, to verify the persistence of the property of interest.

A mapping satisfying a strong validation relation between model and target system indicates therefore the ability of the formal model to fully express the behaviours of relevance in the target system. Such an ability does not require truthfulness, as the system under investigation is required only to have both a formal and a computational model to account for those properties, but no commitment is made on the origin of its content, i.e. whether it expresses some properties and relations actually existing in nature. Strong simulationism expresses the strongest level of representational robustness. As a consequence of our definitions, model robustness is linked to a target system through strong validation:

Thesis 16 (Robustness) *A formal model is a robust representation of a system if it strongly validates it.*

We want now to establish the corresponding maximal criterion at the level of verification, i.e. the relation between implementation and a computational model, assuming robustness of the formal model. In other words, we require for the epistemic analysis of verification an appropriate counterpart notion which preserves the predictive behaviour of the formal model across to the implemented computational artefact. This is expressed in terms of reliability:

Definition 150 (Reliable Computational Artefact) *A computational artefact is reliable if and only if*

- *it correctly implements the appropriate level of abstraction determined by the computational model;*
- *it correctly satisfies all inferential properties of the chosen level of abstraction;*
- *it provides an interpretation of variables resistant to equivalent implementations of the underlying algorithmic translation.*

The problem of obtaining reliable results from constructed entities like models[16] can be reformulated in our framework in view of the persistence of strong relations between

---

[14] This is similar to the formal structures proposed in French and Ladyman (1999), and Bueno (1997).

[15] This is considered in (Weisberg, 2013, ch.9.3.3).

[16] See e.g. Morrison (2015).

target system and formal model (strong validation), between formal and computational model (strong simulationism), and between computational model and implementation (strong verification). The reliability of a computational implementation is by definition different from the materiality of a standard experiment (which can close the gap with the target system by directly reconstructing it in those limited cases where this is possible). In this sense, reliability for computational settings can only be guaranteed by the strongest available of such relations:

Thesis 17 (Reliability) *An implementation is reliable with respect to a computational model if it strongly verifies it.*

On the one hand, robustness establishes the formal model as a benchmarking for the definition of the computational model and, in turn, of any implementation. On the other hand, we have already mentioned that the experimental role of the implemented artefact consists in verifying the model, by its inferential ability in exploring the information space delimited by the model. If this inferential ability does not exceed the conceptual limits of the model, i.e. it does not require a revision of it, we say the model is optimal:

Definition 151 (Optimal Model) *The design of a model is optimal if it maximizes the amount of correct data inferred from any implementation while preserving the appropriate level of abstraction.*

We say that a robust model and a reliable implementation are sufficient for a correct and valid computational experiment.

16.4 Minimal Criteria: Usability and Fitness

If maximal criteria of robustness and reliability establish sufficient conditions for validity and correctness of experimental computational processes, on the other hand necessary conditions provide minimal criteria. These conditions apply in particular in those contexts where a conceptual and ontological precedence of the target system over its models, and in turn over its implementation, is not available, such as in the artificial sciences, e.g. in robotics and network theory.[17] In these contexts, computational experiments are characterized often by the absence of initial observations and a missing initial theory. For example in robotics, certain model-building activities might start from a given task to be resolved, but there is no reality to be known: the model defined is intended to produce a possible solution to the task at hand, and as such it provides a possible interpretation of the model in which that task has to be solved.[18] One proceeds with the selection of the correct LoA and without the possibility of determining its full

---

[17] For a specific epistemological analysis of minimal criteria for simulations in the artificial sciences including usability and fitness, see Primiero (2019b).

[18] This is the experimental setting that has been characterized as exploratory in Schiaffonati (2016) as reported in Section 14.3.

scope, because the system of reference is designed along with the development of the experiment. Hence, to each design choice in the experiment corresponds the selection of a given computational path and behaviour in the model. Accordingly, the full state of the target system is approached from an incomplete one and its relation with the formal system is one of approximation, which at its best can reach a weak validation relation.

To reflect this process, the model at each run of the experiment is characterized as *fit for purpose*:

Definition 152 (Fit-for-Purpose Model) *A formal model is fit for purpose if and only if*

- *it allows to be defined at the chosen level of abstraction the required sets of variables from the implementation;*
- *it correctly preserves all inferential properties of the chosen level of abstraction of the implementation;*
- *it ensures validity of properties of any consistent extension of the implementation (i.e. any extension by other properties that are consistent with the current ones).*

Here, robustness is modified so that the analysis no longer originates in the target system, rather the implementation in a computational artefact constitutes the basis for the definition of the model. This reversed relation between abstract model and implementation establishes fitness of the former in terms of weak or approximate verification by the latter:

Thesis 18 (Fitness) *A formal model is fit for purpose if it is weakly or approximately verified by an implemented artefact.*

From our previous analysis, it follows that a model which only satisfies fitness criteria can only weakly or approximately validate any given target system identified on its basis.

In the context of a weak or approximate verification relation, the implemented artefact does not map fully to paths in the computational model. For this reason, we can only guarantee a minimal level of well-functioning. This notion has already been introduced in our analysis of physical systems and corresponds to usability:

Definition 153 (Usable Computational Artefact) *A computational artefact is usable if and only if*

- *it allows for a consistent experimental analysis of the conceptual space determined by its variables and inferential rules;*
- *it allows the design of a fit model on the basis of the chosen variables and inferential rules.*

The first property simply means that a computational artefact implementing a model is usable if its experimental results are consistent across tests performed within a given class of inputs of the same type. The second property means that the model which can be abstracted from the artefact needs to be fit according to Definition 152, i.e. preserving correctness and consistency. Accordingly, we define usability as the minimal condition for model construction:

Thesis 19 (Usability) *A computational artefact must be (at least) usable to weakly or approximately verify a formal model.*

Fitness and usability as expressed by approximate or weak validation and verification define therefore the minimal criteria for those contexts in which robustness and reliability cannot be said to be obtained, either because robustness analysis fails (but there are still results that can be obtained from the modelling and implementation attempts, e.g. negative results), or because the target system is not fixed and its definition is partly a result of those very operations of implementation design, testing, and model definition on their basis. When the experiment has a minimal role in the definition of the model, the latter should minimize the amount of information it requires from it. In this case we say that the design of the model is efficient:

Definition 154 (Efficient Model) *The design of a model is efficient or calibrated if it minimizes the amount of correct data required for any implementation to be useful, while still preserving the appropriate level of abstraction.*

We say that a usable implementation and a fit-for-purpose model are necessary for a correct and valid computational experiment.

16.5 Experimental Computational Validity

The definition of validity and verification properties and the analysis of minimal and maximal criteria for experimental computing have two aims: first, they allow one to address the debate around the explanatory or predictive power of experiments; second, they permit the establishment of a notion of computational validity for the experimental setting. In both cases, we use the underlying formal relations determined by our variants of simulationism.

Let us start with the first issue: we argue that experiments performed by computational artefacts display explanatory power when they correctly implement a computational model presenting a weak simulation relation to the corresponding formal or mathematical model. Recall from Figure 15.5 that, in this case, there is a mapping between computational artefact and target system. The mapping relation is instantiated bottom-up, i.e. the behaviours manifested in the artefact map to behaviours in the target system. In this case, the artefact instantiates a *how-possibly* explanation of the target system: here the algorithmic implementation does not need to be complete, it does not exclude other possible computational implementations of the same behaviours, and it can include formulations of such behaviours that might only loosely reflect the underlying mechanism of the target system. When, instead, the mapping relation is instantiated top-down, i.e. the behaviours manifested in the target system are mapped to behaviours in the artefact (see Figure 15.6), the latter represents a *how-actually* explanation of the target system: in this case the implementation explains precisely the mechanism of the target system.

In the case of strong simulationism, we indicate with the double-headed arrow between the nodes labelled as artefact and target system, a symmetric mapping relation resulting

How-Possibly Explanation	$CS \leadsto^W MM$
How-Actually Explanation	$MM \leadsto^W CS$
Prediction	$CS \leftrightsquigarrow^S MM$
Exploration	$CS \leadsto^A MM$

Figure 16.7 Correspondences between simulationist relations and modelling types

from the corresponding strong simulation between the computational and the formal model of the target system; see Figure 15.7. On the assumption that such mapping exists, any behaviour shown by the computational artefact must find a counterpart in the target system and if one is already indicated by the formal model of the system, then one can be expected to be implementable. As such, the role of the modelling at hand has a *predictive* nature, in that it will show existing or new or future observations about the system.

In approximate simulationism (see Figure 15.8), there is neither an explanation of nor a prediction about the physical system by the computational description. Instead, we are considering an *explorative* modelling, where the implemented artefact can be used to investigate how certain behaviours of the physical system interact with other behaviours of different systems or to perform modelling without specific targets. We summarize the correspondence between simulationist relations and epistemic properties of modelling in Figure 16.7.

Let us now consider the validity of computations in an experimental setting. Recall that in Section 7.4 the notion of computational validity in the context of the Mathematical Foundation had been formulated through a Principle of Algorithmic Dependence (Principle 5) and a Principle of Encoding Independence (Principle 6): they express the requirement of a linguistic formulation of the input–output relation expressed by the computation, and that the resulting properties are stable under a given equivalence class of such linguistic encodings. On the basis of these two principles, a definition of valid formal computation relies entirely on logical admissibility of the algorithmic encoding and its termination in a state that expresses correctness with respect to specification. In Section 12.3, we extended this notion of validity to account for the analysis of physical computation offered in Part II of this volume: in particular, Principle 12 of Architectural Dependence associates computation with the requirements of a physical architecture along with the linguistic encoding; and Principle 13 of Behavioural Dependence connects the result of the computational process to the selection of a correct and admissible physical infrastructure. The resulting notion of valid physical computation adds to formal validity the requirement that the information processing be functionally correct and efficient, in particular expressed by the ability of working correctly even in the presence of limited failure.

It is now our aim to reconsider validity for the experimental foundation. To do so, we have to reflect first of all the requirement that the implementation which is used to perform experiments aiming at verifying a computational hypothesis has a given correspondence with a computational model, and additionally it has to preserve a minimality principle of usability:

Principle 20 (Realizability Dependence) *An experimental computational process P is the usable and correct implementation of a formal model.*

In other words, the experiment depends both on the model (formal and computational) and the preservation of usability properties in a (correct) physical implementation. This avoids the case of spurious executions of random implementations to be accounted for as appropriate means for computational experiments, as well as malfunctioning processes.

Secondly, the type of relation mapped by the implementation directly influences the result of the computational experiment, and it makes the result of the computational experiment differ. We express this property as relational dependence:

Principle 21 (Relational Dependence) *An experimental computational process P maps differently to the target system depending on the relations it satisfies with its computational and formal models.*

In other words, the results of a computational experiment depend on the types of verification instantiated by the implementation with the computational model, and the simulationist relation of the latter with the formal model.

Finally, we can define validity of an experimental computational process as the conditions for asserting a correct and appropriate mapping between implementation and target system:

Definition 155 (Valid Experimental Computation) *A physical artefact t implements a valid experimental computational process P if and only if*

1. *the formal model MM of P is fit for purpose and robust;*
2. *there is a computational model CM strongly simulating MM;*
3. *the implementation of CM in t is usable, reliable, and it guarantees a valid physical computation.*

This completes our investigation into computational validity, complementing its formal and physical counterparts.

Exercises

Exercise 152 *What is verification for simulation?*

Exercise 153 *What is validation for simulation?*

Exercise 154 *How is weak verification for simulation defined?*

Exercise 155 *How is strong verification for simulation defined?*

Exercise 156 *How is approximate verification for simulation defined?*

Exercise 157 *How is weak validation for simulation defined?*

Exercise 158 *How is strong validation for simulation defined?*

Exercise 159 *How is approximate validation for simulation defined?*

Exercise 160 *Explain the robustness criterion for model definition.*

Exercise 161 *Explain the reliability criterion for simulation.*

Exercise 162 *Explain the optimality criterion for model definition.*

Exercise 163 *Explain the fitness criterion for model definition.*

Exercise 164 *Explain the usability criterion for simulation.*

Exercise 165 *Which type of simulationist relation between computational model and formal model supports how-possibly explanation and why?*

Exercise 166 *Which type of simulationist relation between computational model and formal model supports how-actually explanation and why?*

Exercise 167 *Which type of simulationist relation between computational model and formal model supports prediction and why?*

Exercise 168 *Which type of simulationist relation between computational model and formal model supports exploration and why?*

Exercise 169 *Explain the notion of experimental computational validity and its supporting principles.*

17 Conclusion

Computing is a complex and multi-faceted discipline and our journey through its foundations has unveiled some of its essential aspects: computing as a mathematically grounded
science; computing as an art developing engineering principles; computing as an experimental method. Within each trend, computing includes several issues and problems. But
the most complex topics lie at their intersection. We have offered a reading of algorithms
as layered artefacts, which express abstract rules, linguistically formulated processes, and
implemented routines. We have illustrated how computing machines fail, and how as such
they are prone to lose the perfection of their purely mathematical models. We have considered the nature of computationally produced ontologies as valid only in comparison
to their models, and the need for verification according to scientific principles.

In unveiling these issues we have considered problems which are at the very root of
contemporary society and of the relation between humanity and computing technologies.
Our scientific aim is thus complemented by a social one: this book is a plea for a
methodologically rigorous approach to computing and its overall impact, as well as for
a critical stand towards its ontological and epistemological principles. Once these are
solidly grounded, we believe that ethical, political, and social principles can be formulated
in a similarly consistent fashion.

While the mathematical, formal, and technological principles at the basis of computing
should answer to scientific rigour, the critical stand we encourage should answer to a
philosophical rigour. In this sense, we believe, a pluralistic approach open to the various
aspects of computing is what offers the best option to developing an understanding of
computing in relation to human nature. This volume has not offered a complete analysis as
such: we have not discussed computing from the points of view of the user, of the market,
of the companies, of the legislator, and so on. We have only approached its foundational
aspects, and have left all the remaining complexity to further endeavours. This is a task
that neither scientists nor philosophers alone can hope to pursue successfully. The era
of computing, as any truly new era, requires a new method and a multi-disciplinary
approach.

On the Foundations of Computing. Giuseppe Primiero, Oxford University Press (2020). © Giuseppe Primiero.
DOI: 10.1093/oso/9780198835646.001.0001

Bibliography

Adams, R. *An Early history of Recursive Functions and Computability – From Gödel to Turing*. Docent Press, 2011.

Alberts, G. and E.G. Daylight. Universality versus locality: the Amsterdam style of ALGOL implementation. *IEEE Annals of the History of Computing*, 4:52–63, October–December 2014.

Amdahl, G.M. Validity of the single processor approach to achieving large scale computing capabilities. *Proceedings of AFIPS '67 Spring Joint Computer Conference*, pages 483–5, 1967.

Angius, N. and G. Primiero. The logic of identity and copy for computational artefacts. *Journal of Logic and Computation*, 2018. doi: 10.1093/logcom/exy012. URL https://doi.org/10.1093/logcom/exy012.

Arif, R., E. Mori, and G. Primiero. Validity and correctness before the OS: the case of LEO I and LEO II. In L. De Mol and G. Primiero, editors, *Reflections on Programming Systems – Historical and Philosophical Aspects*, volume 133 of *Philosophical Studies Series*. Springer, 2018.

Arkoudas, K. and S. Bringsjord. Computers, justification, and mathematical knowledge. *Minds and Machines*, 17(2):185–202, 2007.

Ashenhurst, R.L. Letters in the ACM Forum. *Communications of the ACM*, 32:3, 1989.

Aspray, W. The Intel 4004 microprocessor: What constituted invention? *IEEE Annals of the History of Computing*, 19(3):4–15, 1997.

Baarden, J. and W.H. Brattain. Three-electrode circuit element utilizing semiconductive materials. *U.S. Patent*, 2524, February 1948a. URL https://patents.google.com/patent/US2524035A/en.

Baarden, J. and W.H. Brattain. The transistor, a semi-conductor triod. *Phys. Rev*, 74:230–231, 1948b.

Babbage, C. *Babbage's Calculating Engines – Being a Collection of Papers Relating to Them; Their History, and Construction*. Cambridge University Press, New York, 2010. Edited by Babbage, H.P. Original published by A. and F.N. Spon, London, 1889.

Backus, J. Can programming be liberated from the von Neumann style?: A functional style and its algebra of programs. *Commun. ACM*, 21(8):613–641, August 1978. ISSN 0001-0782. doi: 10.1145/359576.359579. URL http://doi.acm.org/10.1145/359576.359579.

Baier, C. and J.P. Katoen. *Principles of Model Checking*. MIT press, 2008.

Barberousse, A., S. Franceschelli, and C. Imbert. Computer simulations as experiments. *Synthese*, 169(3):557–74, 2009.

Barnes, G. R. Brown, M. Kato, D. Kuck, D. Slotnick, and R. Stokes. The ILLIAC IV computer. *IEEE Transactions on Computers*, 17(8):746–57, 1968.

Basili, V.R. The role of experimentation in software engineering: Past, current and future. In *Proc. 18th Int. Conf. Software Eng*, Los Alamitos, Calif, March 1996. IEEE Computer Soc. Press.

Basili, V.R., R.W. Selby, and D.H. Hutchens. Experimentation in software engineering. *IEEE Transactions on Software Engineering*, 12(7):733–43, 1986.

Batini, C. and M. Scannapieco. *Data Quality: Concepts, Methodologies and Techniques*. Springer, New York, 2006.

Battistelli, L. and G. Primiero. Logic-based collective decision making of binary properties in an autonomous multi-agent system. Technical report, Middlesex University London, 06 2017. URL https://www.researchgate.net/publication/327137545_Logic-based_Collective_Decision_Making_of_Binary_Properties_in_an_Autonomous_Multi-Agent_System.

Beggs, E.J. and J.V. Tucker. Can Newtonian systems bounded in space, time, mass and energy compute all functions? *Theoretical Computer Science*, 371(1–2):4–19, 2007.

Böhm, C. and G. Jacopini. Flow diagrams, Turing machines, and languages with only two formation rules. *Communications of the ACM*, 9(5):366–71, 1966.

Bondyopadhyay, P.K. Moore's law governs the silicon revolution. *Proceedings of the IEEE*, 86(1): 78–81, Jan 1998. ISSN 0018-9219. doi: 10.1109/5.658761.

Boolos, G.S. J.P. Burgess, and R.C. Jeffrey. *Computability and Logic*. Cambridge University Press, fourth edition, 2002.

Börger, E. The ASM refinement method. *Formal Asp. Comput.*, 15(2–3):237–57, 2003. doi: 10.1007/s00165-003-0012-7. URL https://doi.org/10.1007/s00165-003-0012-7.

Bottone, M. F. Raimondi, and G. Primiero. Multi-agent based simulations of block-free distributed ledgers. In Leonard Barolli, Makoto Takizawa, Tomoya Enokido, Marek R. Ogiela, Lidia Ogiela, and Nadeem Javaid, editors, *32nd International Conference on Advanced Information Networking and Applications Workshops, AINA 2018 workshops, Krakow, Poland, May 16–18, 2018*, pages 585–90. IEEE Computer Society, 2018. ISBN 978-1-5386-5395-1. doi: 10.1109/WAINA. 2018.00149. URL https://doi.org/10.1109/WAINA.2018.00149.

Bournez, O. A. Pouly, and A. Survey. On analog models of computation. In Vasco Brattka and Peter Hertling, editors, *Handbook of Computability and Complexity in Analysis*. Springer, 2018.

Boyer, R. and Y. Yu. Automated proofs of object code for a widely used microprocessor. *Journal of the ACM*, 43(1):166–92, Jan. 1996.

Bringsjord, S. A vindication of program verification. *History & Philosophy of Logic*, 36(3):262–77, 2015.

Brooks, F. *The Mythical Man-Month*. Addison-Wesley, 1975.

Brooks, F. No silver bullet – essence and accidents of software engineering. *Computer*, 20(4):10–19, April 1987. ISSN 0018-9162. doi: 10.1109/MC.1987.1663532.

Brooks, F. The computer scientist as toolsmith II. *Communications of the ACM*, 39(3):1998, March 1996.

Brouwer, E.L.J. *Over de grondslagen der wiskunde*. PhD thesis, University of Amsterdam, Department of Physics and Mathematics, 1907.

Brouwer, E.L.J. Zur Begründung der intuitionistischen Mathematik I. *Mathematische Annalen*, 93:244–57, 1925a.

Brouwer, E.L.J. Zur Begründung der intuitionistischen Mathematik II. *Mathematische Annalen*, 95:453–72, 1925b.

Brown, J. and J.W. Carr. Automatic programming and and its development on the MIDAC. In *Symposium on Automatic Programming for Digital Computers*, pages 84–97. Office of Naval Research, Department of the Navy, May 1954.

Bueno, O. Empirical adequacy: A partial structures approach. *Studies in History and Philosophy of Science Part A*, 28(4):585 – 610, 1997. ISSN 0039-3681. doi: https://doi.org/10.1016/S0039-3681(97)00012-5. URL http://www.sciencedirect.com/science/article/pii/S0039368197000125.

Bueno, O. S. French, and J. Ladyman. On representing the relationship between the mathematical and the empirical. *Philosophy of Science*, 69(3):497–518, 2002.

Bullynck, M. Reading Gauss in the computer age: on the U.S. reception of Gauss's number theoretical work (1938–1989). *Arch. Hist. Exact Sci*, 63(5):553–80, 2009.

Bullynck, M. Histories of algorithms: Past, present and future. *Historia Mathematica , 43 (3)*, pages 332–41, 2015.

Bullynck, M. What is an operating system? A historical investigation (1954–1964). In L. De Mol and G. Primiero, editors, *Reflections on Programming Systems – Historical and Philosophical Aspects*, Philosophical Studies Series. Springer, 2018.

Bullynck, M. and L. De Mol. Setting-up early computer programs. D.H. Lehmer's ENIAC computation. *Archive for Mathematical Logic*, 49(2):123–46, 2010.

Burali-Forti, C. Una questione sui numeri transfiniti. *Rendiconti del Circolo Matematico di Palermo*, 11:154–64, 1897.

Burge, T. Computer proof, apriori knowledge, and other minds. *Noûs*, 32:1–37, 1998.

Burroughs: Corporation Burroughs. ILLIAC IV Brochure, 1974.

Calcagno, C. and D. Distefano. Infer: An automatic program verifier for memory safety of C programs. In M. Bobaru, K. Havelund, G.J. Holzmann, and R. Joshi, editors, *NASA Formal Methods. NFM 2011*, Lecture Notes in Computer Science. Springer, vol. 6617. Springer, Berlin, Heidelberg, 2011.

Campbell-Kelly, M. *ICL: A Business and Technical History*. Cambridge University Press, New York, NY, 1989. ISBN 0-19-853918-5.

Cantor, G. Über die Ausdehnung eines Satzes aus der Theorie der trigonometrischen Reihen. *Mathematische Annalen*, 5:123–32, 1872. URL http://www.maths.tcd.ie/pub/HistMath/People/Cantor/Ausdehnung/.

Cantwell Smith, B. The limits of correctness. In R. Kling, editor, *Computerization and Controversy*, pages 18–26. ACM SIGCAS Computers and Society, second edition, Morgan Kaufmann, Academic Press, San Diego, 1996.

Carnap, R. Die logizistische Grundlegung der Mathematik. *Erkenntnis*, 2(1):91–105, 1931. Translated as The logicist foundations of mathematics in P. Benaceraff, H. Putnam, editors, *Philosophy of Mathematics – Selected readings*, pp.42–51, Cambridge University Press, 1983.

Carrara, M. and P. Giaretta. Identity criteria and sortal concepts. In *Proceedings of the International Conference on Formal Ontology in Information Systems*, pages 234–43. ACM, 2001.

Cartwright, N. *How the Laws of Physics Lie*. Oxford University Press, 1983.

Ceruzzi, P. *A History of Modern Computing*. MIT Press, second edition, 2003.

Chabert, J.-L. E. Barbin, M. Guillemot, A. Michel-Pajus, J. Borowczyk, and A.D.J.-C. Martzloff. *A History of Algorithms. From the Pebble to the Microchip*. Springer, Berlin, 1999.

Chalmers, D.J. On implementing a computation. *Minds & Machines*, 4:391–402, 1995.

Chrisley, R.L. Why everything doesn't realize every computation. *Minds & Machines*, 4:403–30, 1995.

Church, A. A formulation of the simple theory of types. *Journal of Symbolic Logic*, 5:56–68, 1934.

Church, A. An unsolvable problem of elementary number theory. *American Journal of Mathematics*, 58(2):345–63, 1936.

Churchland, P.S. and T.J. Sejnowski. *The Computational Brain*. MIT Press, Cambridge, MA, 1992.

Clarke, E.M. and J.M. Wing. Formal methods: state of the art and future directions. *ACM Comput. Surv.*, 28(4):1996, December 1996. doi: https://doi.org/10.1145/242223.242257.

Colburn, T. *Philosophy and Computer Science*. M.E. Sharpe, Armonk, NY, 2000.

Colburn, T. and G. Shute. Abstraction in computer science. *Minds & Machines*, 17:169–84, 2007.

Constable, R.L. Constructive mathematics and automatic program writers. In *Proceedings IFIP Congress*, pages 229–233, 1971.

Constable, R.L. Programs as proofs: A synopsis. *Information Processing Letters*, 16:105–12, 1983.

Cook, T.D. and D.T. Campbell. *Quasi-Experimentation – Design and Analysis Issues for Field Settings*. Houghton Mifflin Compa- ny, 1979.

Copeland, B.J. and G. Sommaruga. The stored-program universal computer: Did Zuse anticipate Turing and von Neumann? In G. Sommaruga and T. Strahm, editors, *Turing's Revolution*, pages 43–101. Springer, International Publishing Switzerland, 2015.

Copeland, B.J. and others. *Colossus: The Secrets of Bletchley Park's Codebreaking Computers*. Oxford University Press, 2010.

Cristian, F. Exception handling and software fault-tolerance. *IEEE Transactions on Computers*, 31:531–40, 1982.

Cristian, F. A rigorous approach to fault-tolerant programming. *IEEE Transactions on Software Engineering*, 11:23–31, 1985.

Cristian, F. Understanding fault-tolerant systems. *Communications of the ACM*, 34:56–78, 1991.

Crossley, J.N. and A.S. Henry. Thus spake al-Khwārizmī: A translation of the text of Cambridge University Library ms. ii.vi.5. *Historia Mathematica*, 17(2):103–31, 1990.

Culler, D. J.P. Singh, and A. Gupta. *A Parallel Computer Architecture. A Hardware/Software Approach*. Morgan Kaufmann Publishers Inc., San Francisco, CA, 1998.

Curry, H.B. Functionality in combinatory logic. *Proceedings of the National Academy of Sciences of the United States of America*, 20(11):584–90, 1934.

Curry, H.B. On the composition of programs for automatic computing. *Naval Ordnance Laboratory Memorandum*, 9806(52):19–8, 1949.

da Costa, N.C.A. and S. French. *Science and Partial Truth*. Oxford University Press, 2003.

Davis, M. *The Universal Computer – The Road from Leibniz to Turing*. CRC Press, 2012.

Daylight, E.G. Dijkstra's rallying cry for generalization: the advent of the recursive procedure, late 1950s–early 1960s. *The Computer Journal*, 54(11):1756–72, 2011.

Daylight, E.G. *The Dawn of Software Eingineering*. Lonely Scholar, 2012.

de Beer, H.T. *The History of the ALGOL Effort*. Technische Universiteit Eindhoven, Department of Mathematics and Computer Science, 2006.

de Bruijn, N.G. A survey of the project AUTOMATH. In *Symposium on Automatic Demonstration*, volume 125 of *Lecture Notes in Mathematics*, pages 29–61, 1970.

de Bruijn, N.G. A survey of the project automath. In J.R. Hindley J.P. Seldin, editors, *To H.B. Curry: Essays on Combinatory Logic, Lambda Calculus and Formalism*, pages 579–606. Academic Press, 1980.

De Millo, R.L., R.J. Lipton, and A.J. Perlis. Social processes and proofs of theorems and programs. *Communications of the ACM*, 22(5):271–81, 1979.

De Mol, L. Turing machines. In Edward N. Zalta, editor, *The Stanford Encyclopedia of Philosophy*. Metaphysics Research Lab, Stanford University, winter 2018 edition, 2018.

De Mol, L. and G. Primiero. Facing computing as technique: Towards a history and philosophy of computing. *Philosophy & Technology*, 27(3):321–26, 2014.

De Mol, L. and G. Primiero. When logic meets engineering: Introduction to logical issues in the history and philosophy of computer science. *History & Philosophy of Logic*, 36(3):195–204, 2015.

De Mol, L., M. Carlé, and M. Bullynck. Haskell before Haskell: an alternative lesson in practical logics of the ENIAC. *J. Log. Comput*, 25(4):1011–46, 2015.

Dean, W. Algorithms and the mathematical foundations of computer science. In L. Horsten and P. Welch, editors, *The Limits of Mathematical Knowledge*. Oxford University Press, 2016.

Dedekind, R. *Was sind und was sollen die Zahlen?* Braunschweig, Friederich Beiweg and Son, 1888.

Dedekind, R. *Stetigkeit und Irrationale Zahlen*. Braunschweig, Friederich Beiweg and Son, 1872. Republished by Springer, 1960.

Denning, P.J. Performance evaluation: experimental computer science at its best. *ACM Performance Evaluation Review*, 1981:106–9, 1981.

Denning, P.J. Computer science. In *Encyclopedia of Computer Science*, pages 405–19. John Wiley and Sons Ltd., Chichester, UK, 2000. ISBN 0-470-86412-5. URL http://dl.acm.org/citation.cfm?id=1074100.1074266.

Denning, P.J. Is computer science science? *Communications of the ACM*, 48:27–31, April 2005.

Denning, P.J. and T.G. Lewis. Exponential laws of computing growth. *Communications of the ACM*, 60(1):54–65, 2017. doi: 10.1145/2976758.

Denning, P.J., E. Feigenbaum, P. Gilmore, A. Hearn, R.W. Ritchie, and J. Traub. A discipline in crisis. *Communications of the ACM*, 24(6):1981, June 1981.

Denning, P.J., D.E. Comer, D. Gries, M.C. Mulder, A. Tucker, J. Turner, and P.R. Young. Computing as a discipline. *Communications of the ACM*, 32:9–23, January 1989.

Dershowitz, N. and Y. Gurevich. A natural axiomatization of computability and proof of Church's thesis. *Bulletin of Symbolic Logic*, 14(3):299–350, 2008. URL http://www.math.ucla.edu/%7Easl/bsl/1403/1403-002.ps.

Deutsch, D. Quantum theory, the Church-Turing principle and the universal quantum computer. *Proceedings of the Royal Society of London A*, 400:97–117, 1985.

Dijkstra, E.W. The humble programmer. *Communications of the ACM*, 15(10):859–66, October 1972.

Dijkstra, E.W. Programming as a discipline of mathematical nature. *American Mathematical Monthly*, 81:608–612, 1974.

Dijkstra, E.W. Correctness concerns and, among other things, why they are resented. In *Proceedings of the 1975 International Conference on Reliable Software*, pages 546–50. ACM SIGPLAN Notices, Los Angeles, California U.S.A, April 1975.

Dijkstra, E.W. *A Discipline of Programming*. Prentice-Hall, 1976.

Dijkstra, E.W., Duncan, Garwick, Hoare, Randell, Seegmueller, Turski, and Woodger. Minority report. Technical Report 31, Working Group 2.1, March 1970.

Durán, J. *Computer Simulations in Science and Engineering – Concepts, Practices, Perspectives*. Springer, 2018.

Durán, J.M. A brief overview of the philosophical study of computer simulations. *American Philosophical Association Newsletter on Philosophy and Computers*, 13(1):38–46, 2013.

Eckert, J.P. A preview of a digital computing machine. In C.C. Chambers, editor, *Theory and Techniques for the Design of Electronic Digital Computers*, pages 1–10. Moore School of Electrical Engineering, University of Pennsylvania, pp.10–26, 1947.

Eckert, J.P. A survey of digital computer memory systems. *Proceedings of the I.R.E*, 1406(1953): 15–28, 1998.

Eden, A.H. Three paradigms of computer science. *Minds & Machines*, 17(2):135–67, July 2007. URL http://dx.doi.org/10.1007/s11023-007-9060-8.

Eins: The Eins Consortium. Overview of ICT energy consumption (d8.1). Technical report, European Network of Excellence in Internet Science, February 2013.

Ensmenger, N. *The Computer Boys Take Over – Computers, Programmers, and the Politcs of Technical Expertise*. The MIT Press, Cambridge Massachusetts, London England, 2010.

Faggin, F. The making of the first microprocessor. *IEEE Solid-State Circuit Magazine*, 1(1):8–21, 2009.

Faggin, F. The MOS silicon gate technology and the first microprocessors. *La Rivista del Nuovo Cimento*, 38(12):575–621, 2015.

Fant, K.M. Critical review of the notion of algorithm in computer science. In *Proceedings of the 1993 ACM Conference on Computer Science (CSC '93). ACM, NY, USA*, pages 1–6, 1993.

Feldman, J., and W.R. Sutherland. Rejuveneting experimental computer science – a report to the National Science Foundation and others. *Communications of the ACM*, 22(9):1979, 1979.

Fetzer, J.H. Program verification: The very idea. *Communications of the ACM*, 31(9):1048–63, 1988.

Floridi, L. *The Philosophy of Information*. Oxford University Press, 2011.

Floridi, L. The method of abstraction. In L. Floridi, editor, *The Routledge Handbook of Philosophy of Information*, pages 50–6. Routledge, 2014.

Floridi, L. The logic of design as a conceptual logic of information. *Minds & Machines*, 27(3): 495–519, Sep 2017. URL https://doi.org/10.1007/s11023-017-9438-1.

Floridi, L., N. Fresco, and G. Primiero. On malfunctioning software. *Synthese*, 192(4):1199–220, 2015.

Floyd, R.W. Assigning meanings to programs. *Proceedings of Symposia in Applied Mathematics*, 19: 19–32, 1967.

Fodor, J.A. *The Language of Thought*. Harvard University Press, Cambridge, MA, 1975.

Forsythe, G. A university's educational program in computer science. *Communications of the ACM*, 10(1):3–11, 1967.

Forsythe, G. What to do till the computer scientist comes. *American Mathematical Monthly*, 75: 454–61, 1968.

Van Fraassen, B.C. *The Scientific Image*. Oxford University Press, 1980.

Frege, G. *Grundgesetze der Arithmetik*, volume I. Verlag Hermann Pohle, Jena, 1903. URL http://www.korpora.org/Frege/.

Frege, G. Concept script, a formal language of pure thought modelled upon that of arithmetic. In Jan van Heijenoort, editor, *From Frege to Gödel: A Source Book in Mathematical Logic*, pages 1879–1931. Harvard University Press, Cambridge, MA, 1967.

French, S. and J. Ladyman. Reinflating the semantic approach. *International Studies in the Philosophy of Science*, 13(2):103–21, 1999.

Fresco, N. Explaining computation without semantics: Keeping it simple. *Minds and Machines*, 20: 165–81, 2010.

Fresco, N. *Physical Computation and Cognitive Science*. Springer, New York, 2014.

Fresco, N. and G. Primiero. Miscomputation. *Philosophy & Technology*, 26(3):253–72, 2013.

Frigg, R. and J. Reiss. The philosophy of simulation: hot new issues or same old stew? *Synthese*, 169 (3):593–613, 2009.

Gandy, R. Church's thesis and principles for mechanisms. In Jon Barwise, H. Jerome Keisler, and Kenneth Kunen, editors, *The Kleene Symposium*, volume 101 of *Studies in Logic and the Foundations of Mathematics*, pages 123–48. Elsevier, 1980. doi: https://doi.org/ 10.1016/S0049-237X(08)71257-6. URL http://www.sciencedirect.com/science/article/pii/ S0049237X08712576.

Gandy, R. The confluence of ideas in 1936. In *A Half-century Survey on The Universal Turing Machine*, pages 55–111, New York, NY, USA, 1988. Oxford University Press, Inc. ISBN 0-19-853741-7. URL http://dl.acm.org/citation.cfm?id=57249.57252.

Gelenbe, E. and Y. Caseau. The impact of information technology on energy consumption and carbon emissions. *Ubiquity*, 2015, June 2015.

Gelenbe, E. and R. Lent. Energy-QoS trade-offs in mobile service selection. *Future Internet*, 5(2): 128–39, 2013.

Gelenbe, E. and C.A Morfopoulou. Framework for energy-aware routing in packet networks. *Comput. J*, 54(6):850–9, 2011.

Génova, G. Is computer science truly scientific? *Communications of the ACM*, 53(7):37–9, July 2010.

Gentzen, G. Untersuchungen über das logische Schließen. I. *Mathematische Zeitschrift*, 39(1): 176–210, Dec 1935. URL https://doi.org/10.1007/BF01201353.

Gilbert, G.N. and K.G. Troitzsch. *Simulation for the Social Scientist*. Taylor & Francis, Inc. Bristol, PA, USA, 1999.

Gillespie, M. *Amdahl's Law, Gustafson's Trend, and the Performance Limits of Parallel Applications*. Intel, 2008a. URL https://software.intel.com/sites/default/files/m/d/4/1/d/8/Gillespie-0053-AAD_Gustafson-Amdahl_v1__2_.rh.final.pdf.

Gillespie, M. *Scaling Software Architectures for the Future of Multi-Core Computing*. Intel, 2008b. URL http://www.dell.com/downloads/global/power/ps4q08-50090144-Intel-SW.pdf.

Girard, J.-Y. Linear logic. *Theoretical Computer Science*, pages 1–102, 1987.

Girard, J.-Y. P. Taylor, and Y. Lafont. *Proof and Types*. Cambridge University Press, Cambridge, 1989.

Gödel, K. Über die Vollständigkeit des Logikkalküls (1929). Reprinted in Collected Works, Volume I, Publications 1929–1936, by Kurt Gödel, edited by Feferman Solomon, Dawson John W. Jr., Kleene Stephen C., Moore Gregory H., Solovay Robert M., and van Heijenoort Jean, as Gödel, K. *On the Completeness of the Calculus of Logic*. English Translation by Stefan Bauer-Mengelberg and Jean van Heijenoort. Clarendon Press, Oxford University Press, New York and Oxford 1986, pp. 60–100.

Gödel, K. Über formal unentscheidbare Sätze der Principia Mathematica und verwandter Systeme, I. *Monatshefte für Mathematik und Physik*, 38:144–95, 1931.

Gödel, K. On formally undecidable propositions of Principia Mathematica and related systems i. In Solomon Feferman, editor, *Kurt Gödel Collected Works*, volume I, pages 173–98. Oxford University Press, 1986.

Godfrey-Smith, P. Triviality arguments against functionalism. *Philosophical Studies*, 145(2): 273–95, 2009.

Goldstine, H.H. *The Computer. From Pascal to von Neumann*. Princeton University Press, Princeton, 1972.

Goldstine, H.H. *A History of Numerical Analysis. From the 16th Through the 19th Century*. Springer, New York, Heidelberg, 1977.

Goldstine, H.H. and J. von Neumann. Planning and coding of problems for an electronic computing instrument. Technical Report, Institute of Advanced Studies, Princeton, 1947.

Goodenough, J. Exception handling: issues and a proposed notation. *Commun. ACM*, 8:683–96, 1975.

Goodman, N. *Fact, Fiction, and Forecast*. Bobbs-Merrill, Indianapolis, 1965.

Gordon, M. HOL: A proof generating system for higher-order logic. In *VLSI Specification, Verification and Synthesis*. Kluwer, 1987.

Gorn, S. Planning universal semi-automatic coding. In *Symposium on Automatic Programming for Digital Computers, Office of Naval Research, Department of the Navy, , D.C., pp.74–83*, pages 13–14, May 1954.

Gorn, S. Standardized programming methods and universal coding. *Journal of the ACM*, 1957.

Gorn, S. Common programming language task: Final report. *AD*, 59:1–201, 1959.

Govindarajan, R. Exception handlers in functional programming languages. *IEEE Transactions on Software Engineering*, 19:826–34, 1993.

Grantsröm, J.G. *Treatise on Intuitionistic Type Theory*. Springer, 2011.

Grassmann, H. *Lehrbuch der Arithmetik für höhere Lehranstalten*. T.F. Enslin, Berlin, 1861. URL https://archive.org/details/lehrbuchderarit00grasgoog.

Grochowski, A. A. Bhattacharya, D. Viswanathan, and T.R. Laker. Integrated circuit testing for quality assurance in manufacturing: History, current status, and future trends. *IEEE Transactions On Circuits and Systems - II : Analog and Digital Signal Processing*, 44:61–633, 1997.

Guala, F. Models, simulations, and experiments. In *Model-based reasoning. Science, technology, values*, pages 59–74. Springer, 2002.

Guala, F. Paradigmatic experiments: The ultimatum game from testing to measurement device. *Philosophy of Science*, 75:658–69, 2008.

Gurevich, Y. Evolving algebras 1993: Lipari guide. In E. Börger, editor, *Specification and validation methods*, pages 9–36. Oxford University Press, 1993. ISBN 0-19-853854-5.

Gurevich, Y. Sequential abstract-state machines capture sequential algorithms. *ACM Tansactions on Comutational Logic*, 1:77–111, 2000.

Gurevich, Y. What is an algorithm? In M. Bieliková, G. Friedrich, G. Gottlob, S. Katzenbeisser, and G. Turán, editors, *SOFSEM 2012: Theory and Practice of Computer Science – 38th Conference on Current Trends in Theory and Practice of Computer Science, Špindlerův Mlýn, Czech Republic, 21–27 January 2012. Proceedings*, volume 7147 of *Lecture Notes in Computer Science*, pages 31–42. Springer, 2012. doi: 10.1007/978-3-642-27660-6. URL https://doi.org/10.1007/978-3-642-27660-6_3.

Gustafson, J.L. Reevaluating Amdahl's law. *Communications of the ACM*, 31(5):532–3, 1988. doi: 10.1145/42411.42415.

Haigh, T. Inventing information systems: the systems men and the computer. *Business History Review*, 75(1):15–61, 2001.

Haigh, T. Dijkstra's crisis: The end of Algol and beginning of software engineering. Draft for discussion in SOFT-EU Project Meeting, September 2010. URL http://www.tomandmaria.com/Tom/Writing/DijkstrasCrisis_LeidenDRAFT.pdf.

Haigh, T. Stored program concept considered harmful – history and historiography. In Benedikt Löwe, Paola Bonizzoni, Vasco Brattka, editors, *The Nature of Computation. Logic, Algorithms, Applications*, volume 7921 of *Lecture Notes in Computer Science*, pages 241–51. Springer, 2013.

Haigh, T., M. Priestley, and C. Rope. Reconsidering the stored-program concept. *IEEE Annals of the History of Computing*, 36:4–17, 2014.

Haigh, T., M. Pristley, and C. Rope. *ENIAC in Action*. MIT Press, 2016.

Hamming, R.W. One man's view of computer science. *Journal of the ACM*, 16(1):3–12, 1968.

Harel, D. On folk theorems. *Communications of the ACM*, 23(7):379–88, 1980.

Hartmanis, J. On computational complexity and the nature of computer science. *Communications of the ACM*, 37(10):37–43, October 1994.

Hartmanis, J. and H. Lin. What is computer science and engineering? In J. Hartmanis and H. Lin, editors, *Computing the Future: A Broader Agenda for Computer Science and Engineering*, pages 163–216. National Academy Press, Washington, DC, 1992.

Hartmann, S. The world as a process. In *Modelling and Simulation in the Social Sciences from the Philosophy of Science Point of View*, pages 77–100. Springer, 1996.

Hempel, C. *Aspects of Scientific Explanation and Other Essays in the Philosophy of Science*. Free Press, New York–London, 1965.

Hempel, C. *Philosophy of Natural Science*. Prentice-Hall, Englewood Cliffs, 1966.

Herraiz, I., D. Rodriguez, G. Robles, and J.M. Gonzalez-Barahona. The evolution of the laws of software evolution: a discussion based on a systematic literature review. *ACM Computing Surveys*, 46(2):1–28, 2013.

Hevnes, A.R. and D.J. Berndt. Eras of business computing. *Advances in Computers*, 52:1–90, 2000.

Heyting, A. Die formalen Regeln der intuitionistischen Logik. *Sitzungsberichte der Preussischen Akademie von Wissenschaften*, pages 42–56, 1930.

Heyting, A. Die intuitionistische Grundlegung der Mathematik. *Erkenntnis*, 2(1):106–115, 1931. Translated as The intuitionist foundations of mathematics in P. Benacerraf, H. Putnam, editors, *Philosophy of Mathematics – Selected readings*, pp.52–61, Cambridge University Press, 1983.

Hicks, M. *Programmed Inequality: How Britain Discarded Women Technologists and Lost its Edge in Computing*. MIT Press, 2017.

Hilbert, D. *Grundlagen der Geometrie*. Verlag u. Druck v. B.G. Teubner, 1902. URL http://www.gutenberg.org/files/17384/.

Hilbert, D. Über das Unendliche. In Putnam Benacerraf, editor, *Mathematische Annalen*, pages 161–90. *Philosophy of Mathematics*, pp.183–206, 1983.

Hill, R. What an algorithm is. *Philosophy & Technology*, 29(1):35–59, 2016.

Hindley, J.R. and F. Cardone. History of Lambda-calculus and Combinatory Logic. In Dov M. Gabbay and John Woods, editors, *Handbook of the History of Logic*, volume 5. Elsevier Co., 2006.

Hoare, C.A.R. An axiomatic basis for computer programming. *Communications of the ACM*, 12(10): 576–80, 1969.

Hoare, C.A.R. Mathematics of programming – mathematical laws help programmers control the complexity of tasks. In Timothy R. Colburn, Terry L. Rankin, and James H. Fetzer, editors, *Program Verification: Fundamental Issues in Computer Science*, pages 135–54. Kluwer, 1993.

Hoare, C.A.R. How did software get so reliable without proof? In M.-C. Gaudel and J. Woodcock, editors, *FME'96: Industrial Benefit and Advances in Formal Methods*. FME 1996. Lecture Notes in Computer Science, vol. 1051. Springer, Berlin, Heidelberg, 1996.

Hornbeck, J.A. The transistor. In F.M Smits, editor, *A History of Engineering and Science in the Bell System: Electronics Technology and The Transistor (1925–1975)*, pages pp.1–100. AT & T Bell Laboratories, 1985.

Howard, W.A. The formulae-as-types notion of construction. In J.P. Seldin and J.R. Hendlin, editors, *Essays on Combinatory Logic, Lambda Calculus, and Formalism*, pages 479–91. Academic Press, NY, 1980.

Hubig, C. and A. Kaminski. Outlines of a pragmatic theory of truth and error in computer simulation. In Michael Resch, Andreas Kaminski, and Petra Gehring, editors, *Science and Art of Simulation I*, pages 121–36. Springer, Berlin, Heidelberg, 2017.

Humphreys, P. Computer simulations. *PSA: Proceedings of the Biennial Meeting of the Philosophy of Science Association*, 1990:497–506, 1990.

Humphreys, P. *Extending Ourselves: Computational Science, Empiricism, and Scientific Method*. Oxford University Press, 2004.

Humphreys, P. The philosophical novelty of computer simulation methods. *Synthese*, 169(3): 615–26, 2009.

Huth, M. and M. Ryan. *Logic in Computer Science*. Cambridge University Press, second edition, 2004.

Illari, P. IQ: Purpose and dimensions. In P. Illari L. Floridi, editor, *The Philosophy of Information Quality*, pages 281–301. Synthese Library vol. 358, Springer, 2014.

Imrack, N. Software is an abstract artifact. *Grazer Philosophische Studien*, 86(1):55–72, 2012.

International Energy Agency. 2012 Key World Energy Statistics. Technical report, OECD/IEA, 2012. URL http://alofatuvalu.tv/FR/12_liens/12_articles_rapports/IEA_rpt_2012_us.pdf.

Ishtiaq, S.S. and P.W. O'Hearn. BI as an assertion language for mutable data structures. *ACM SIGPLAN Notices*, 36(3):14–26, 2001.

Israeli, A.F. The Linux kernel as a case study in software evolution. *J. Syst. Software*, 83(3):485–501, 2010.

Jones, C.B. *Systematic Software Development Using VDM*. Prentice-Hall International, New York, 1986.

Jones, C.B. The early search for tractable ways of reasoning about programs. *IEEE Annals of the History of Computing*, 25(2):26–49, 2003.

Jones, C.B. Turing's 1949 paper in context. In Jarkko Kari, Florin Manea, and Ion Petre, editors, *Unveiling Dynamics and Complexity*, pages 12–16. 32–41, 13th Conference on Computability in Europe, CiE 2017, Turku, Finland, June *Lecture Notes in Computer Science*, 10307, 2017.

Kaminski, A., M. Resch, and U. Küster. Mathematische Opazität. Über Rechfertigung und Reproduzierbarkeit in der Computersimulation. In A. Friedrich, P. Gehring, C. Hubig, Kaminski A., and A. Nordmann, editors, *Jahrbuch Technikphilosophie*, pages 253–78. Nomos Verlagsgesellschaft, 2018.

Kaner, C., J. Bach, and B. Pettichord. *Lessons Learned in Software Testing*. John Wiley & Sons, Inc., New York, NY, 2001. ISBN 0471081124.

Kennedy, H.C. *Selected Works of Giuseppe Peano*. University of Toronto Press, 1973. ISBN 9781487592240. URL http://www.jstor.org/stable/10.3138/j.ctt1vxmd8x.

Khazzoom, J.D. Economic implications of mandated efficiency standards for household appliances. *The Energy Journal*, 1(4):21–40, 1980.

Kilby, J.S. Miniaturized electronic circuits [US Patent no. 3,138, 743]. *IEEE Solid-State Circuits Society Newsletter*, 12(2):44–54, Spring 2007. ISSN 1098-4232. doi: 10.1109/N-SSC.2007. 4785580.

Kitcher, P. Explanatory unification and the causal structure of the world. In P. Kitcher and W. Salmon, editors, *Scientific Explanation*, pages 410–505. Minneapolis: University of Minnesota Press, 1989.

Kleene, S.C. A theory of positive integers in formal logic. *American Journal of Mathematics*, 57: 153–73, 1935.

Kleene, S.C. General recursive functions on natural numbers. *Mathematische Annalen*, 112:727–42, 1936a.

Kleene, S.C. A note on recursive functions. *Bulletin of the American Mathematical Society*, 42:544–6, 1936b.

Kleene, S.C. λ-definability and recursiveness. *Duke Mathematical Journal*, 2:340–53, 1936c.

Kleene, S.C. *Introduction to Metamathematics*. North-Holland, 1952.

Klein, T. and F. Faggin. Doped semiconductor electrodes for MOS type devices. *US Patent*, 3, June 1972.

Knuth, D. *The Art of Computer Programming*. Addison-Wesley, Reading, Massachusetts, third edition, 1997.

Knuth, D.E. Ancient babylonian algorithms. *Communications of the ACM*, 15(7):671–7, 1972.

Knuth, D.E. Computer science and and its relation to mathematics. *American Mathematical Monthly*, 81(4):323–43, 1974.

Knuth, D.E. Algorithms in modern mathematics and computer science. In Andrei P. Ershov and Donald E. Knuth, editors, *Algorithms in Modern Mathematics and Computer Science*, volume 122 of *Lecture Notes in Computer Science*, pages 82–99. Springer, Berlin, Heidelberg, 1981.

Knuth, D.E. Algorithmic thinking and mathematical thinking. *The American Mathematical Monthly*, 92(3):170–81, 1985. URL http://www.jstor.org/stable/2322871.

Kolmogorov, A. Zur Deutung der intuitionistischen Logik. *Mathematische Zeitschrift*, 35:58–65, 1932.

Koomey, J.G. Growth in data center electricity use 2005 to 2010. *Oakland, CA, Analytics Press*, 1, August 2011.

Korb, K.B. and S. Mascaro. The philosophy of computer simulation. In G. Glymour, W. Wei, and D. Westerstahl, editors, *Logic, Methodology and Philosophy of Science: Proceedings of the Thirteenth International Congress*, pages 306–25. College Publications, 2009.

Kripke, S.A. The Church-Turing thesis as a special corollary of Gödel's completeness theorem. In B.J. Copeland, C. Posy, and O. Shagrir, editors, *Computability: Turing, Gödel, Church, and Beyond*. MIT Press, 2013.

Kuehlmann, A., A. Srinivasan, and D.P. LaPotin. Verity – a formal verification program for custom CMOS circuits. *IBM J. Res. Dev*, 39(1):149–65, 1995.

Land, F. The first business computer: a case study in user-driven innovation. *IEEE Annals of the History of Computing*, 22(3):16–26, 2000.

Lavington, S.H. Manchester computer architectures, 1948–75. *IEEE Annals of the History of Computing*, 15(3):44–54, 1993.

Lee, E.A. Absolutely positively on time. *IEEE Computer*, pages 85–87, July 2005.

Lee, Y.W., D.M. Strong, B.K. Kahn, and R.Y. Wang. AIMQ: A methodology for information quality assessment. *Information & Management*, 40(2):133–46, 2002.

Lehman, M.M. On understanding laws, evolution, and conservation in the large-program life cycle. *Journal of Systems and Software*, 1:213–21, 1979.

Lehman, M.M. Programs, life cycles and laws of software evolution. *Proceedings of the IEEE*, 68: 1060–76, 1980.

Lehman, M.M. Laws of software evolution revisited. In *Proceedings of the European Workshop on Software Process Technology*, pages 108–24. Springer-Verlag, London, 1996.

Lehman, M.M. and L.A. Belady. *Program evolution. Processes of software change*. Academic Press, San Diego, CA, 1985.

Lehman, M.M. and J. Fernández-Ramil. Software evolution. In N.H. Madhavji, J. Fernández-Ramil, and D.E. Perry, editors, *Software Evolution and Feedback. Theory and Practice*, pages 7–40. Wiley, 2006.

Leibniz, G.W. *Philosophical Papers and Letters - A Selection*, volume 2 of *Synthese Historical Library*. Springer, 1989.

Lindsey, C.H. A history of ALGOL 68. In Thomas J. Bergin and Rick G. Gibson, editors, *History of Programming Languages II*, pages 27–83. ACM Press, New York, 1996.

Llull, R. *Selected Works*. Princeton University Press, 1985. Edited by Anthony Bonner.

Loui, M.C. Computer science is a new engineering discipline. *ACM Computing Surveys*, 27(1):1995, March 1995.

Lowe, E.J. Objects and criteria of identity. In B. Hale and C. Wright, editors, *A Companion to the Philosophy of Language*, pages 990–1012. Blackwell Publishers Ltd, Oxford, 1997.

Luo, Z. and R. Pollack. Lego proof development system: User's manual. Technical report, Computer Science Dept., Univ. of Edinburgh, May 1992.

Mack, C.A. Fifty years of Moore's Law. *IEEE Transactions on Semiconductor Manufacturing*, 24(2): 202–7, 2011.

Manna, Z. The correctness of programs. *Journal of Computer and System Sciences*, 3:119–27, 1969.

Manna, Z. Mathematical theory of partial correctness. *Journal of Computer and System Sciences*, 5: 239–53, 1971.

Manna, Z. and J. McCarthy. Properties of programs and partial function logic. *Machine Intelligence*, 5:79–98, 1969.

Martin-Löf, P. Constructive mathematics and computer programming. In *Proceedings of the Sixth International Congress for Logic, Methodology and Philosophy of Science*, pages 153–75, 1982.

Martin-Löf, P. *Intuitionistic Type Theory*. Bibliopolis, Napoli, 1984.

Marwedel, P. and M. Engel. Efficient computing in cyber physical systems. In *2012 International Conference on Embedded Computer Systems (SAMOS)*, pages 328–32. Samos, 2012.

Maydanchik, A. *Data Quality Assessment*. Technics Publications, LLC, Bradley Beach NJ, 2007.

McClusky, E.J. and S. Mitra. Fault tolerance. In A.B. Tucker, editor, *Computer Science Handbook*, chapter 25. CRC Press, second edition, 2004.

McKee, G. Computer science or simply 'computics'? *Computer*, 28(12):136, December 1995. ISSN 0018-9162. doi: 10.1109/MC.1995.476210. URL https://doi.org/10.1109/MC.1995.476210.

Milkowski, M. *Explaining the Computational Mind*. MIT Press, Cambridge, MA, 2013.

Milne, R. and C. Strachey. *A Theory of Programming Language Semantics*. Halsted Press, New York, NY, 99th edition, 1977. ISBN 0470989068.

Milner, R. An algebraic definition of simulation between programs. Technical Report, CS-205, Department of Computer Science, Stanford University, 1971.

Mollick, E. Establishing Moore's Law. *IEEE Annals of the History of Computing*, 28(3):2006, July–Sept. 2006.

Moor, J.H. Three myths of computer science. *The British Journal for the Philosophy of Science*, 29(3): 213–22, 1978.

Moore, G.E. Cramming more components onto integrated circuits. *Electronics*, 117(19):82–5, April 1965.

Moore, G.E. Progress in digital integrated electronics. *IEDM Tech. Digest*, 1975:11–13, 1975.

Moore, G.E. Lithography and the future of Moore's Law. *Proceedings SPIE*, 2437:1995, May 1995.

Morgan, M. Experiments without material intervention: model experiments, virtual experiments and virtually experiments. In H. Radder, editor, *The Philosophy of Scientific Experimentation*, pages 216–35. University of Pittsburgh Press, Pittsburgh, PA, 2003.

Morrison, M. Models, measurement and computer simulation: the changing face of experimentation. *Philosophical Studies*, 143(1):33–57, 2009.

Morrison, M. *Reconstructing Reality: Models, Mathematics, and Simulations*. Oxford University Press, USA, 2015.

Morse, S.F.B. Improvement in the mode of communicating information by signals by the application of electro-magnetism. *US Patent*, 1647, June 1840.

Moschovakis, Y.N. What is an algorithm? In B. Engquist and W. Schmid, editors, *Mathematics Unlimited – 2001 and Beyond*, pages 919–36. Springer, 2001 (Part II). 24 April 2001.

Mosses, P.D. Formal semantics of programming languages: an overview. *Electronic Notes in Theoretical Computer Science*, 148(1):41–73, 2006. URL http://www.sciencedirect.com/science/article/pii/S1571066106000429. *Proceedings of the School of SegraVis Research Training Network on Foundations of Visual Modelling Techniques* (FoVMT 2004).

Mowery, D.C. 50 years of business computing: LEO to Linux. *The Journal of Strategic Information Systems*, 12(4):295–308, 2003.

Naur, P. and B. Randell, editors. *Software Engineering: Report of a Conference Sponsored by the NATO Science Committee, Garmisch, Germany, 7–11 Oct. 1968*. Scientific Affairs Division, NATO, 1969.

Newell, A., A.J. Herbert Perlis, and A. Simon. What is computer science? *Science*, 157:1373–4, 1967.

Nofre, D. Unravelling ALGOL: U.S., Europe, and the creation of a programming language. *IEEE Annals of the History of Computing*, 32(2):58–68, 2010.

Nordström, B., K. Petersson, and J.M. Smith. *Programming in Martin-Löf's Type Theory*. Oxford University Press, 1990.

North, M.J. and C.M. Macal. *Managing Business Complexity: Discovering Strategic Solutions with Agent-based Modeling and Simulation*. Oxford University Press, 2007.

Oddifreddi, P. *Classical Recursion Theory – The Theory of Functions and Sets of Natural Numbers*. North-Holland, Amsterdam-London-New York-Tokyo, 1992.

O'Hearn, P.W. and D.J. Pym. The logic of bunched implications. *Bulletin of Symbolic Logic*, 5:02, 1999.

O'Regan, F. *A Brief History of Computing*. Springer, 2012.

Parker, W.S. Does matter really matter? Computer simulations, experiments and materiality. *Synthese*, 169(3):483–96, 2009.

Parker, W.S. Computer simulation. In S. Psillos and M. Curd, editors, *The Routledge Companion to Philosophy of Science*, pages 135–45. Routledge, 2013.

Paruthi, V. Large-scale application of formal verification: From fiction to fact. *FMCAD*, pages 175–80, 2010.

Paulson, L.C. Constructing recursion operators in Intuitionistic Type Theory. *Journal of Symbolic Computation*, 2(4):325–55, 1986.

Peano, G. *Arithmetices prinicipia, novo methodo exposita*. Bocca, Turin, 1889. URL https://archive.org/details/arithmeticespri00peangoog.

Peirce, C.S. On the logic of number. *American Journal of Mathematics*, 4(1):85–95, 1881.

Piccinini, G. *Physical Computation: A Mechanistic Account*. Oxford University Press, 2015, Oxford, 2015.

Piccinini, G. Computation in physical systems. In *The Stanford Encyclopedia of Philosophy*. ward N. Zalta (ed.), 2017. URL https://plato.stanford.edu/archives/sum2017/entries/computation-physicalsystems/.

Piccinini, G. and N.G. Anderson. Ontic pancomputationalism. In M.C. Cuffaro and S.C. Fletcher, editors, *Physical Perspectives on Computation, Computational Perspectives on Physics*, chapter 1. Cambridge University Press, 2018.

Pickavet, M. Worldwide energy needs for ICT: the rise of power-aware networking. In *2008 Second International Symposium on Advanced Networks and Telecommunication Systems*, pages 1–3. Mumbai, 2008.

Plaice, J. Computer science is an experimental science. *ACM Computing Surveys*, 27:1, 1995.

Poincaré, H. La logique de l'infini. *Revue de Métaphysique et de Morale*, 17(4):461–82, 1909.

Popper, K.R. *The Logic of Scientific Discovery*. Routledge, 2002, London, 1959.

Post, E.L. Finite combinatory processes – formulation 1. *Journal of Symbolic Logic*, 1(3):103–05, 09, 1936. URL https://projecteuclid.org:443/euclid.jsl/1183142134.

Prawitz, D. *Natural Deduction*. Almqvist and Wiksell, Stockholm, 1965.

Price, R. Oral history interview of Shannon, C.E. In *Interview #423 for the IEEE History Center*, volume 28. The Institute of Electrical and Electronics Engineers, July 1982.

Priestley, M. *A Science of Operations – Machines, Logic and the Invention of Programming*. History of Computing. Springer, 2011. ISBN 978-1-84882-554-3. doi: 10.1007/978-1-84882-555-0. URL https://doi.org/10.1007/978-1-84882-555-0.

Primiero, G. *Information and Knowledge – A Constructive Type-Theoretical Approach*. Springer, 2006.

Primiero, G. An epistemic logic for becoming informed. *Synthese*, 167(2):363–89, 2009. URL https://doi.org/10.1007/s11229-008-9413-8.

Primiero, G. Offline and online data: on upgrading information to knowledge. *Philosophical Studies*, 164(2):371–392, 2013.

Primiero, G. A taxonomy of errors for information systems. *Minds and Machines*, 24(3):249–73, 2014.

Primiero, G. Realist consequence, epistemic inference, computational correctness. In A. Koslow and A. Buchsbaum, editors, *The Road to Universal Logic, vol. II*, pages 573–88. Studies in Universal Logic, Springer International Publishing, Switzerland, 2015.

Primiero, G. Information in the philosophy of computer science. In L. Floridi, editor, *The Routledge Handbook of Philosophy of Information*, pages 90–106. Routledge, 2016.

Primiero, G. Algorithmic iteration for computational intelligence. *Minds and Machines*, 27(3): 521–43, 2017. doi: 10.1007/s11023-017-9423-8. URL https://doi.org/10.1007/s11023-017-9423-8.

Primiero, G. A logic of efficient and optimal designs. *Journal of Logic and Computation*, 2019a. doi: 10.1093/logcom/exz014.

Primiero, G. A minimalist epistemology for agent-based simulations in the artificial sciences. *Minds & Machines*, 29(1):127–148, 2019b. URL https://doi.org/10.1007/s11023-019-09489-4.

Primiero, G., F. Raimondi, M. Bottone, and J. Tagliabue. Trust and distrust in contradictory information transmission. *Applied Network Science*, 2:12, 2017a. URL https://doi.org/10.1007/s41109-017-0029-0.

Primiero, G., F. Raimondi, T. Chen, and R. Nagarajan. A proof-theoretic trust and reputation model for VANET. In *2017 IEEE European Symposium on Security and Privacy Workshops, EuroS&P Workshops 2017, Paris, France, 26–28 April 2017*, pages 146–52. IEEE, 2017b. ISBN 978-1-5386-2244-5. URL https://doi.org/10.1109/EuroSPW.2017.64.

Primiero, G., A. Martorana, and J. Tagliabue. Simulation of a trust and reputation based mitigation protocol for a black hole style attack on VANETs. In *2018 IEEE European Symposium on Security and Privacy Workshops, EuroS&P Workshops 2018, London, United Kingdom, 23–27 April 2018*, pages 127–35. IEEE, 2018a. ISBN 978-1-5386-5445-3. URL https://doi.org/10.1109/EuroSPW.2018.00025.

Primiero, G., F.J. Solheim, and J. Spring. On malfunction, mechanisms, and malware classification. *Philosophy & Technology*, 2018b. URL https://doi.org/10.1007/s13347-018-0334-2.

Primiero, G., E. Tuci, J. Tagliabue, and E. Ferrante. Swarm attack: A self-organized model to recover from malicious communication manipulation in a swarm of simple simulated agents. In Marco Dorigo, Mauro Birattari, Christian Blum, Anders Lyhne Christensen, Andreagiovanni Reina, and Vito Trianni, editors, *Swarm Intelligence – 11th International Conference, ANTS 2018, Rome, Italy, 29–31 October 2018, Proceedings*, volume 11172 of *Lecture Notes in Computer Science*, pages 213–24. Springer, 2018c. ISBN 978-3-030-00532-0. doi: 10.1007/978-3-030-00533-7. URL https://doi.org/10.1007/978-3-030-00533-7_17.

Putnam, H. Minds and machines. In S. Hook, editor, *Dimensions of Mind: A Symposium*, pages 138–64. Cambridge University Press, Cambridge, 1975.

Pylyshyn, Z.W. *Computation and Cognition*. MIT Press, Cambridge, MA, 1984.

Pym, D., J.M. Spring, and P. O'Hearn. Why separation logic works. *Philosophy & Technology*, 32(3):483–516, 2019. URL https://doi.org/10.1007/s13347-018-0312-8.

Quine, W.V.O. *Pursuit of truth*. Harvard University Press, 1990.

Rajcham, J.A. Magnetic system. *U.S. Patent*, 2792, May 1957. URL https://patents.google.com/patent/US2792563/.

Ramsey, F.P. *The Foundations of Mathematics*. Routledge, 1926.

Randell, B. The origins of computer programming. *IEEE Annals of the History of Computing*, 16: 6–14, 1994.

Rapaport, W.J. Implementation is semantic interpretation. *The Monist*, 82(1):109–30, 1999.

Rapaport, W.J. Implementation is semantic interpretation: Further thoughts. *Journal of Experimental and Theoretical Artificial Intelligence*, 17(4):385–417, 2005.

Rapaport, W.J. Semiotic systems, computers, and the mind: how cognition could be computing. *International Journal of Signs and Semiotic Systems*, 2(1):32–71, 2012.

Rapaport. W.J. *Philosophy of Computer Science*. Draft of May, 2018.

Ratliff. T.A. *The Laboratory Quality Assurance System – A Manual of Quality Procedures and Forms*. Wiley, Interscience, third edition, 2003.

Reichenbach, M. and R.S. Cohen. The Königsberg conference on the epistemology of the exact sciences [1930f]. In Maria Reichenbach and Robert S. Cohen, editors, *Hans Reichenbach Selected Writings 1909–1953: Volume One*, pages 324–5. Springer Netherlands, Dordrecht, 1978. ISBN 978-94-009-9761-5. doi: 10.1007/978-94-009-9761-5_36. URL https://doi.org/10.1007/978-94-009-9761-5_36.

Rennels, D. Fault-tolerant computing – concepts and examples. *IEEE Transactions on Computers*, 33:1116–29, 1984.

Reus, B. *Limits of Computation: From a Programming Perspective*. Springer Publishing Company, first edition, 2016.

Reynolds, J.C. Separation logic: a logic for shared mutable data structures. In *Proceedings of 17th annual IEEE Symposium on Logic in Computer Science*, pages 55–74, 2002.

Riordan, M. and L. Hoddeson. *Crystal Fire: The Birth of the Information Age*. Norton, 1997.

Rojas, R. How to make Zuse's Z3 a universal computer. *IEEE Annals of the History of Computing*, 20 (3):51–4, 1998.

Rojas, R. and U. Hashagen. *The First Computers – History and Architectures*. MIT Press, 2002.

Ross, I.M. The invention of the transistor. *Proceedings of the IEEE*, 86(1):7–28, 1998.

Rubin, J. and D. Chisnell. *Handbook of Usability Testing: How to Plan, Design, and Conduct Effective Tests*. Wiley Publishing, second edition, 2008.

Russell, B. *The Principles of Mathematics*. Cambridge University Press, Cambridge, 1903.

Russell. B. Mathematical logic as based on the theory of types. *American Journal of Mathematics*, 1908:59–102, 1956.

Russell. B. Letter to Frege, 26th June 1902. In J. van Heijenoort, editor, *From Frege to Gödel: A Sourcebook in Mathematical Logic*, pages 124–5. Harvard University Press, 1965.

Sakellari, G., C. Morfopoulou, T. Mahmoodi, and E. Gelenbe. Using energy criteria to admit flows in a wired network. *ISCIS 2012*, pages 63–72, 2013.

Schaller, R.R. Moore's law: past, present and future. *IEEE Spectrum*, 34(6):52–9, June 1997. doi: 10.1109/6.591665.

Scheutz, M. When physical systems realize functions.... *Minds & Machines*, 9(2):161–96, 1999.

Schiaffonati, V. Stretching the traditional notion of experiment in computing: explorative experiments. *Science and Engineering Ethics*, 22(3):647–65, June 2016.

Schorr, H. Experimental computer science. In *Proceedings of a symposium on Computer culture: the scientific, intellectual, and social impact of the computer*, pages 31–46, New York Academy of Sciences, New York, 1984. New York Academy of Sciences.

Schubert, K.D. Solutions to IBM POWER8 verification challenges. *IBM Journal of Research and Development*, 59(1):1–11, 2015.

Scott, D.S. and C. Strachey. Toward a mathematical semantics for computer languages. In *Proc. Symp. on Computers and Automata*, volume 21 of *Microwave Research Institute Symposia Series*. Polytechnic Institute of Brooklyn, 1971.

Shagrir, O. Why we view the brain as a computer. *Synthese*, 153(3):393–416, 2006.

Shannon, C.E. A symbolic analysis of relay and switching circuits. *Transactions American Institute of Electrical Engineers*, 57:471–95, 1938.

Shannon, C.E. *A Universal Turing Machine with two internal states*. In Shannon and McCarthy (1956), pp.157–65, 1956.

Shannon, C.E. and J. McCarthy. *Automata Studies*. Princeton University Press, 1956.

Shmueli, G. To explain or to predict? *Statistical Science*, 25(3):289–310, 2010. URL http://www.jstor.org/stable/41058949.

Sieg, W. Calculations by Man & Machine: Conceptual analysis. In W. Sieg, R. Sommer, and C. Talcott, editors, *Reflections on the Foundations of Mathematics*, pages 396–415. Association for Symbolic Logic, 2002.

Sieg, W. Church without dogma. In B. Löwe, S.B. Cooper, and A. Sorbi, editors, *New Computational Paradigms: Changing Conceptions of What Is Computable*. Springer, LNCSNew York; London, 2008.

Siewert, S. Big iron lessons, part 2: Reliability and availability: what's the difference? Technical report, IBM, 03 2005.

Skolem, T. Begründung der elementaren Arithmetik durch die rekurriende Denkweise ohne Anwendung scheinbarer Veränglichen mit unendlichen Ausdehnungsbereich. *Vidensk. Skrifter*, 6:303–33, 1923.

Smith, W.D. Church's thesis meets the n-body problem. *Applied Mathematics and Computation*, 178 (1):154–83, 2006.

Sørensen, M.H. and P. Urzyczyn. *Lectures on the Curry-Howard Isomorphism*. Elsevier, 2006.

Spielmann, M. Automatic verification of abstract state machines. In N. Halbwachs and D.A. Peled, editors, *Computer Aided Verification, 11th International Conference, CAV '99, Trento, Italy, 6–10 July 1999, Proceedings*, volume 1633 of *Lecture Notes in Computer Science*, pages 431–42. Springer, 1999. URL https://doi.org/10.1007/3-540-48683-6_37.

Spielmann, M. Model checking abstract state machines and beyond. In Y. Gurevich, P.W. Kutter, M. Odersky, and L. Thiele, editors, *Abstract State Machines, Theory and Applications, International Workshop, ASM 2000, Monte Verità, Switzerland, 19–24 March 2000, Proceedings*, volume 1912 of *Lecture Notes in Computer Science*, pages 323–40. Springer, 2000. URL https://doi.org/10.1007/3-540-44518-8_18.

Spivey, J.M. *Introducing Z: A Specification Language and its Formal Semantics*. Cambridge University Press, New York, 1988.

Suber, P. What is software? *Journal of Speculative Philosophy*, 2(2):89–119, 1988.

Sun, S., J.-L. Bertrand-Krajewski, A. Lynggaard-Jensen, J. van den Broeke, F. Edthofer, M. do Céu Almeida, A. Silva Ribeiro, and J. Menaia. Literature review of data validation methods. Technical report, – Seventh Framework Programme, 2011.

Tal, E. From data to phenomena and back again: computer-simulated signatures. *Synthese*, 182(1): 117–29, 2011.

Tarski, A. The semantic conception of truth and the foundations of semantics. *Philosophy and Phenomenological Research*, 4(3):341–76, 1943.

Teal, G.K. Single crystals of germanium and silicon – basic to the transistor and integrated circuit. *IEEE Transactions on Electron Devices*, 23:621–39, 1976.

Team: The Coq Development Team. The coq proof assistant reference manual. *v.*, 8:7, 2017. URL https://coq.inria.fr/distrib/current/files/Reference-Manual.pdf.

Tedre, M. Computing as engineering. *Journal of Universal Computer Science*, 15(8):1642–58, 2009.

Tedre, M. *The Science of Computing – Shaping a Discipline*. CRC Press, 2015.

Tedre, M. and N. Moisseinen. Experiments in computing: a survey. *The Scientific World Journal*, 2013. doi: http://dx.doi.org/10.1155/2014/549398.

Tennent, R.D. The denotational semantics of programming languages. *Commun. ACM*, 19(8): 437–53, August 1976. ISSN 0001-0782. URL http://doi.acm.org/10.1145/360303.360308.

Thompson, S. *Type Theory and Functional Programming*. Addison-Wesley, 1991.

Tichy, W.F. Should computer scientists experiment more? *Computer*, 31(5):32–40, May 1998.

Tomayko, J.E. Computers in spaceflight: the NASA experience. Technical report, NASA, 1988.

Trenholme, R. Analog simulation. *Philosophy of Science*, 61(1):115–31, 1994. URL https://doi.org/10.1086/289783.

Troelstra, A.S. *Principles of Intuitionism*, volume 95 of *Lecture Notes in Mathematics*. Springer Verlag, 1969.

Turing, A.M. On computable numbers, with an application to the Entscheidungsproblem. *Proceedings of the London Mathematical Society*, 2(42):230–65, 1936.

Turing, A.M. Checking a large routine. In *Report of a Conference on High Speed Automatic Calculating Machines*, pages 67–9, 1949. Republished in F.L. Morris and C.B. Jones, Annals of the History of Computing, 6(2).

Turing, A.M. *Programmers' Handbook for the Manchester Electronic Computer Mark I*, 1951. URL http://curation.cs.manchester.ac.uk/computer50/www.computer50.org/kgill/mark1/RobertTau/turing.pdf.

Turner, R. Machines. In H. Zenil, editor, *A Computable Universe: Understanding and Exploring Nature as Computation*, pages 63–76. World Scientific Publishing Company/Imperial College Press, London, 2012.

Turner, R. *Computational Artifacts – Towards a Philosophy of Computer Science*. Springer, 2018. URL https://doi.org/10.1007/978-3-662-55565-1.

Turner, R. and N. Angius. The philosophy of computer science. In N. Zalta, editor, *The Stanford Encyclopedia of Philosophy (Spring 2017 Edition)*. Metaphysics Research Lab, Stanford University, 2017. URL https://plato.stanford.edu/archives/spr2017/entries/computer-science/.

Tweddle, J.C. Weierstrass's construction of the irrational numbers. *Mathematische Semesterberichte*, 58(1):47–58, 2011.

Tymoczko, T. The four-color problem and its philosophical significance. *The Journal of Philosophy*, 76(2):57–83, 1979.

Tymoczko, T. Computers, proofs and mathematicians: a philosophical investigation of the four-color proof. *Mathematics Magazine*, 53(3):131–8, 1980.

Van Heddeghem, W., S. Lambert, B. Lanoo, D. Colle, M. Pickavet, and P. Demeester. Trends in worldwide ICT electricity consumption from 2007 to 2012. *Computer Communications*, 50: 64–76, 2014.

Varenne, F. The nature of computational things – models and simulations in design and architecture. In M.A. Brayer and F. Migayrou, editors, *Naturalizing Architecture*, pages 96–105. Hyx editions, 2013.

von Neumann, J. First draft of a report on the EDVAC. *Contract No, 670*, June 1945.

von Neumann, J. Die formalistische Grundlegung der Mathematik. *Erkenntnis* 2(1):116–121, 1931. Translated as The formalist foundations of mathematics in P. Benacerraf, H. Putnam, editors, *Philosophy of Mathematics - Selected readings*, pp.61–65, Cambridge University Press, 1983.

Wadler, P. Propositions as types. *Communications of the ACM*, 58(12):75–84, December 2015.

Wallace, C.S. *Statistical and Inductive Inference by Minimum Message Length (Information Science and Statistics)*. Springer-Verlag, Berlin, Heidelberg, 2005.

Wang, R.Y. and D.M. Strong. Beyond accuracy: what data quality means to data consumers. *Journal of Management Information Systems*, 12:4, 1996.

Wang, R.Y.A. Product perspective on total data quality management. *Communications of the ACM*, 41(2):58–65, 1998. doi: 10.1145/269012.269022.

Wang, X., N. Guarino, G. Guizzardi, and J. Mylopoulos. Towards an ontology of software: a requirements engineering perspective. *FOIS 2014*, pages 317–29, 2014.

Weisberg, M. *Simulation and Similarity – Using Models to Understand the World*. Oxford University Press, 2013.

Whitehead, A.N. and B. Russell. *Principia Mathematica, 3 vols*. Cambridge University Press, Cambridge, second edition, 1962.

Wiggins, D. *Sameness and Substance Renewed*. Cambridge University Press, 2001.

Winsberg E. A tale of two methods. *Synthese*, 169(3):575–92, 2009.

Winsberg, E. *Science in the Age of Computer Simulation*. University of Chicago Press, Chicago, IL, 2010.

Winsberg, E. Computer simulations in science. In Edward N. Zalta, editor, *The Stanford Encyclopedia of Philosophy (Summer Edition)*. Metaphysics Research Lab, Stanford University, 2015. URL http://plato.stanford.edu/archives/sum2015/entries/simulations-science/.

Wohlin, C., P. Runeson, M. Bost, M.C. Ohlsson, B. Regnell, and A. Wesslén. *Experimentation in Software Engineering – An Introduction*. Springer, New York, 2000.

Woodcock, J., P.G. Larsen, J. Bicarregui, and J. Fitzgerald. Formal methods: practice and experience. *ACM Comput. Surv.*, 41(4):19:1–19:36, October 2009. ISSN 0360-0300. doi: 10.1145/1592434.1592436. URL http://doi.acm.org/10.1145/1592434.1592436.

Woodward, J. *Making Things Happen: A Theory of Causal Explanation*. Oxford University Press, 2003.

Woolfson, M.M. and G.J. Pert. *An Introduction to Computer Simulation*. Oxford University Press, 1999.

Yang, H. and P. O'Hearn. A semantic basis for local reasoning. In *FoSSaCS '02 Proceedings of the Fifth International Conference on Foundations of Software Science and Computation Structures*, Lecture Notes in Computer Science, pages 402–16, Springer, Berlin, 2002.

Index